Food Webs

MONOGRAPHS IN POPULATION BIOLOGY
EDITED BY SIMON A. LEVIN AND HENRY S. HORN

A complete series list follows the index

Food Webs

KEVIN S. MCCANN

PRINCETON UNIVERSITY PRESS
Princeton and Oxford

Library of Congress Cataloging-in-Publication Data

McCann, Kevin S. (Kevin Shear), 1964–
    Food webs / Kevin S. McCann.
        p. cm. — (Monographs in population biology)
    Includes bibliographical references and index.
ISBN 978-0-691-13417-8 (hardback) -- ISBN 978-0-691-13418-5 (paperback)
1. Food chains (Ecology) 2. Biotic communities. I. Title.
QH541.M23 2011
577'.16—dc23   2011019434

British Library Cataloging-in-Publication Data is available

This book has been composed in Times New Roman

Printed on acid-free paper. ∞

Typeset by S R Nova Pvt Ltd, Bangalore, India

Printed in the United States of America

10 9 8 7 6 5 4 3 2 1

*If species composing a particular ecosystem begin to go extinct,
at what point will the whole machine sputter and destabilize?*

—E. O. Wilson, *The Diversity of Life* (1992)

Contents

PART 3

TOWARD WHOLE SYSTEMS

Preface

I started thinking about this book after being approached by Sam Elworthy, who suggested a book based on recent food web theory, a synthesis of sorts. At about the same time, Joe Rasmussen, a friend and colleague at McGill told me that he believed no one understands recent food web theory. Taken together, I felt there was a need for such a book and that a synthesis with more attention to conceptual details was clearly in order. One of the more active research areas of late pertains to the "theory of weak interactions," which I and numerous collaborators have played a role in developing. While this theory emphasizes weak interactions, it necessarily also considers the role of interactions of all strengths. This book seeks to illustrate how interaction strength governs the dynamics of food webs. As such, it is a very broadly based synthesis.

Nonetheless, the book does not cover all areas of food webs exhaustively nor does it attempt to. Further, as a monograph, it necessarily pays attention to work developed both in my laboratory and with my many excellent collaborators. At the end of each chapter I briefly scan empirical evidence. In some cases, this constitutes a review or meta-analysis, but frequently it is really just a cursory look at the emerging data and is done to promote further comparison among theory, experiment, and empirical patterns. Clearly this book is not a full review, as to provide such a review would be well beyond its scope.

The ideas laid out in this book are best developed with the language of mathematics, specifically dynamical systems. Thus, not surprisingly, this book has a fair amount of mathematics in it. However, I have made attempts to present it in an accessible format by constantly interpreting all mathematical logic within a biological framework. Further, I have tried to present the intuitive side of the results. You will see that much of this biological interpretation employs a bioenergetic framework. It just as easily could have been interpreted in terms of limiting nutrients or other relevant currencies, but this was the language of my Ph.D. advisor, Peter Yodzis, and I have grown accustomed to it.

For those who understand mathematics, I hope this book will be easy, and for those who do not, I hope they can harness the biological intuition behind these results and so contribute to the development of further food web theory (either theoretically or empirically). At a certain level, theory is about the development of heightened logical intuition, and so whether one understands the mathematics or not is not as important as understanding the underlying concept. It just so happens that some people find it easier to think about things in terms of x's and y's, and others in terms of rabbits and lynx. I am convinced that all theory in ecology is potentially accessible to those with a more empirical background; however, I think both theoreticians and empiricists must make an effort to overcome the barriers imposed by history. I find this an exciting time, as it is becoming obvious that researchers of both ilk seem quite ready to delve deeply into each other's realms and benefit from each other's different perspectives and knowledge. Biocomplexity initiatives have worked wonders to force collaborations over traditional scientific boundaries, making today's ecological and evolutionary science a truly dynamical research environment.

I owe a great debt to Joe Rasmussen, who has talked to me frequently about various and sundry ideas, but much of this talk has attempted to better train me in ecosystem ecology. I also owe a great debt to my graduate and post-doctoral researchers, James Umbanhowar, Tim Bartley, Gabriel Gellner, Tyler Tunney, Jason Rip, and Colette Ward, who have discussed various of these ideas with me in earlier manifestations. Often one learns a lot from the new ideas that their less focused approach allows. Gabriel Gellner deserves special mention, as he has pushed me to consider community matrices all over again, and many of the ideas related to this comes largely from his pushing me along this axis. He led me to rediscover the excellent work of Daniel Haydon, who employed Gershgorin discs to interpret matrix food web theory—a highly informative and, in my opinion, undercited set of novel contributions. Neil Rooney, a friend and postdoctoral researcher, played a large conceptual role in development of the later chapters. This work originates historically from my early training from Peter Yodzis, Don DeAngelis, and Alan Hastings, all of whom have had an enormous influence on my work—they have collectively formed my scientific perspective, and I feel extremely lucky to have worked with such great mentors. I obviously would not have been able to put this book together without the patience and understanding of my wife, Amy, and my children, Caden, Bronwyn, and Robyn. Amy allows me the freedom to pursue all my interests and does so without question. I thank my parents, without whose support I would never have been in a position to write a book like this in the first place. From day one they had no issue with my pursuing something that seemed unimaginable to them ("You are getting a Ph.D. in zoology. They have careers for that?"). Finally, I am grateful to the people who ultimately support scientific research. I am truly blessed to have the opportunity to ponder daily the workings of one of nature's most beautiful entities, the food web.

PART 1
THE PROBLEM AND THE APPROACH

CHAPTER ONE

The Balance of Nature: What Is It and Why Care?

1.1 BALANCING A NOISY SYSTEM

Each spring as the sun begins to strengthen again, I walk the trail that surrounds our house. Unfailingly, I am met by the steady green carpet of plants, the chorus of songbirds, the scurrying of squirrels, and the occasional hawk presiding over the forest floor. On any spring night I may find myself awakened by the unmistakable cry of coyotes to find that the night is alive with the peeping and buzzing of frogs and insects. At a very rough observational level, the main groups of players that make up this localized food web (i.e., plants, herbivores, predators) appear to be consistently present from year to year. If these observations are correct, we can say that this system is stable in the sense that the species assemblage persists intact over ecological time scales. We as casual observers have grown to expect this consistency. Much of this book is concerned with this aspect of stability, often called persistence in the ecological literature. I would argue that this persistence-based notion of stability has fascinated humans throughout history precisely because they have casually observed this pattern for such a long time and depended upon it for their survival.

Despite the pleasing notion that the world's ecosystems harbor a great steadiness or a perfect balance, the more detailed observer is uncomfortable with this statement. Most biologists today would in fact be quick to argue that if we can expect anything in ecological systems, we can expect change (Levin, 2003). As an example, if we look closer at the lush green carpet of plants, we may find there are far more goldenrods this year than last year. Upon further inspection we may also find that last year's summer storms, more frequent than normal, knocked down many large silver maples, leaving behind forest gaps and impressive new understory growth. Thus, the level below the apparent consistency harbors considerable variation within the plant species themselves. This same phenomenom is true for animals. Some years may see enormous insect outbreaks, followed several years later by the increased presence of insectivorous species that during breeding capitalized on the pulse of abundant prey. Simply put, ecosystems are dynamic entities, waxing and waning at a variety of temporal and spatial scales. As we focus in on them, this beautiful, endless, dynamic mosaic appears to be everywhere, and yet, amazingly, the net result at a more macroscopic level (e.g., that of the casual walking observer) is a complex system that harbors some degree of stasis (i.e., a similar assemblage).

There is something calming to the human mind about this consistency in species assemblage, and there is also something unsettling when things do change dramatically in an ecosystem. In her 1962 book, *Silent Spring*, Rachel Carson (1962) documented the loss of birds to DDT. "Silent spring" referred to the fact that our human expectation of nature (i.e., the sound of birds in spring) had been abruptly altered. Although many challenged her scientific assertions, the book and its ideas endured, trumpeting in the modern environmental movement. Carson's ideas were tangible in that both scientists and lay people were able to observe the loss of a major group of species as a result of a human activity. In a sense, the consistency of our forests, our parks, and our backyards had been threatened.

This loose, persistence-based definition of stability is at the heart of most of the more mathematically rigorous definitions that ecologists have historically employed. Variability, or the coefficient of variation, CV (variation/mean), is a common measure of stability in both experiments and recent theory. The logical argument behind this metric of stability is that the more variable a species' population dynamics are, the more likely that species is to attain dangerously low densities. High variability, all else equal, therefore implies a greater risk of local extinction.

Similarly, resilience, so common to many theoretical equilibrium studies, was argued to be an appropriate measure of stability because a resilient population rapidly returned to near equilibrium densities after a perturbation (Pimm et al., 1991). Thus, if a species is perturbed to low densities, a rapid return time to equilibrium means that this same species quickly rises away from near-zero densities and, in doing so, avoids the threat of local extinction. A slowly returning species, on the other hand, is more likely to be subjected to the vagaries of nature's noisy world for a much greater time and so has a significantly higher risk of extinction. So even mathematically based definitions, which assume equilibrium, are in a real sense attempting to speak to the consistency of a species assemblage in a variable world. Clearly though, theory needs to further explore this casual assumption, and many researchers have started to look more rigorously at the fascinating interaction between environmental variability and population stability [e.g., see Ives and Carpenter (2007); Ives et al. (2008)].

There is a long history of ecologists seeking to understand what factors contribute to the stability of ecological communities. Early ecologists pointed to the role diversity plays in stability. This idea remains to some degree today, but most researchers now seek a more explicit understanding of the mechanisms behind stability. It seems likely that if diversity does truly correlate with stability, this is not because of diversity per se but rather because of some fundamental structures embedded in diversity itself (May, 1974b). This biological structure can be at the population scale (e.g., age structure), the community scale (e.g., food web structure), or the ecosystem scale (e.g., size of detrital compartment). The task remains to uncover these fundamental natural structures—a difficult task for sure because the balancing act of nature couples interactions over an enormous range of spatial scales. At local scales (e.g., 1-m^2 plots), nature's balance seems amiss, with organisms varying in number in both space and time. At large enough scales, local variance may in fact cancel itself out to become a flatlined equilibrium process. Local variability can beget regional stability.

Along these lines and within a single trophic level context, Tilman and others (Tilman et al., 1998; Doak et al., 1998) have recognized that species level variation, under various sets of conditions, can ultimately lead to relatively constant competitive communities. Variation at the plant population level, for example, can sum to give a relatively constant plant community as long as not all species increase or decrease together. These researchers, in a sense, changed the stability question by embracing population level variation in density and focusing on the implications of population variability for whole–plant community stability. Once aware of this complex mosaic of spatial and temporal variance, it becomes interesting to consider how this variation itself may play a role in the stability and sustainability of ecosystems. As such, variation in space and/or time may be considered a form of natural structure that organisms have adapted to thrive within. This book will argue frequently that this may indeed be the case. I will also further argue that most human impacts tend to restructure the landscape with broad, homogenizing strokes. Species loss aside, such actions may remove the intricate, detailed spatial and temporal structure that underlies most pristine ecosystems.

This precise aspect of an ecosystem—the multilayered complex of interacting organisms that transcends small to large spatial and temporal scales—is also the toughest part to study. One can ignore interactions by focusing on an isolated box (e.g., population ecology), yet these scale-dependent connections cannot be easily ignored for large ecological problems like ecosystem stability and function. When ecologists, for example, purposefully separate this scale dependency in controlled microcosm experiments, such as simplified and spatially restricted laboratory universes, these mini-ecosystems often fail rapidly after a few violent oscillations in population dynamics. In aquatic microcosms these spatially simplified worlds almost always end up dominated by bacteria. This experimental result may speak to the notion that ecological systems are enormously dependent on the interconnections that span huge ranges in spatial scale. If so, human actions that leave behind fragmented and less spatially connected ecosystems ought to put ecosystems at grave risk of collapse.

In summary, there appears to be a balance of nature, but it is highly unlikely that we are talking about a system in equilibrium. Rather, the persistence of highly diverse complex ecological systems is an emergent property of an intensely interactive and variable underlying dynamical system. I would argue that ecologists never saw the balance of nature as a perfect equilibrium process and that to attack the concept interpreted as such is to take down a straw man. Within this more generalized definition of balance, it remains an important task to ask what it is about nature that allows it to maintain itself and how these complex natural entities adapt in the face of such a variable world. These scientific tasks are closely aligned with the applied societal need to understand how human modifications will impact the diversity, sustainability, and functioning of ecological systems.

This book is an attempt to conceptually synthesize our current understanding of one of the big questions in ecology and evolution—What governs the stability of ecological systems? Although we have briefly discussed stability above, it is obviously critical to more rigorously define what we mean by stability. In the remaining sections of this chapter, I first define stability, discuss the role whole systems play in governing stable ecosystem function, and punctuate the stability problem with case studies of examples of instability in ecological systems. This final aspect of the chapter is included to convince the reader that there are already numerous examples of ecological instability and ecosystem collapse.

1.2   ECOSYSTEM STABILITY AND SUSTAINABILITY

It is commonly asserted that different definitions of stability often lead to different answers about what governs nature's stability (Ives and Carpenter, 2007). As an example, although Ives and Carpenter (2007) found differences between a number of stability definitions in terms of whether diversity begets stability, they also found that all definitions that involved dynamics consistently gave the same qualitative answer. I will show throughout this book that dynamical definitions of stability are often consistent, and when they are not, it is informative to consider why, as suggested by Ives and Carpenter (2007). For instance, attributes that are stable in one sense (e.g., a population return forms a large perturbation rapidly) may be destabilizing in some other important sense (e.g., the same rapidly returning population overshoots the equilibrium and oscillates). I will show that this dynamic trade-off (i.e., fast return–big overshoot) is common in both population and consumer-resource models and that this result is useful in developing a synthetic theory for stability in more complex food web models.

I now define some common measures of stability. These definitions require some understanding of common terms, such as "equilibrium", used in theory. For those having diffulculty with terminology, it may be worth reading the mathematical review of chapter 2 before reading the stability definitions below.

1.2.1 PERSISTENCE-BASED METRICS OF STABILITY

Much of the theory I will discuss in this book relies on the following broad group of stability metrics (Pimm, 1982, 1984; McCann et al., 2000). I am referring to this set of metrics as persistence-based because they are an attempt, through various means, to quantify how likely the system as a whole is able to persist intact. Persistence-based definitions of stability effectively assume that the underlying dynamical system (i.e., an n-member mathematical model) is not changed by a perturbation (i.e., perturbation does not remove a species).

1.2.1.1 *Engineering Resilience*

A measure that assumes system stability increases with decreasing return time to an attracting state (e.g., equilibrium) after a perturbation. The faster the return time, the more stable the system. In this book the term "resilience" will refer to engineering resilience.

1.2.1.2 *Equilibrium Resilience*

The state being returned to after a perturbation is an equilibrium attractor. Mathematically, it is measured by the inverse of the maximum eigenvalue (i.e., $1/\lambda_{max}$).

1.2.1.3 *Nonequilibrium Resilience*

The state being returned to after a perturbation is a nonequilibrium attractor (e.g., a limit cycle, a chaotic attractor).

1.2.1.4 *Variance Stability*

The variance in population densities over time, usually measured as the coefficient of variation, CV (variance/mean). High variance implies a greater chance of extinction, especially with external perturbations which are common in experimental tests of stability.

1.2.1.5 *Bounded or Minima Stability*

A system is more stable than another system if its global minimum density is bounded further away from zero densities than that of the other system. Here, "bounded further away from zero" simply implies that the minimum population density is further away from zero and so less likely to go extinct in a variable environment or after a perturbation.

1.2.1.6 *Sustainability*

A system is said to be sustainable if its' component members are able to persist in the face of a specified perturbation [either a continuous perturbation (i.e., a press perturbation) or a discrete perturbation (i.e., a pulse perturbation)].

1.2.2 CHANGE-BASED METRICS OF STABILITY

Persistence-based notions of stability, in some sense, all concern themselves with the likelihood that the n-member system persists over ecological time scales. There are also metrics of stability that assume a system can change. With this assumption, stability metrics become more concerned with quantifying how a perturbation changes the system and the extent of the change. This has become very popular with the realization that ecosystems may flip from one attractor to another after a perturbation [called an *alternative state* (Scheffer et al., 2001)]. Some common stability metrics of this sort are the following (Holling, 1996).

1.2.2.1 *Alternative State Stability I*

A system is deemed less stable the more alternative states it has. All else equal, the more alternative states, the greater the likelihood a given pertubation will flip the system to an entirely different state.

This focus on multiple attractors has led Holling to define a related alternative state–based measure of stability (Holling, 1996).

1.2.2.2 *Alternative State Stability II (Holling's Resilience)*

A measure of the amount of change or disruption (i.e., the size of the perturbation) that is required to transform a system from being maintained by one set of mutually reinforcing processes and structures (a given attractor) to a different set of processes and structures (another attractor with potentially different species).

1.2.2.3 *Resistance Stability*

Resistance is a metric that quantifies the change of a system after a perturbation. The smaller the change of the system after the perturbation, the more resistant the system. This metric is commonly used when we consider a perturbation that changes the structure of the system, such as the complete removal or addition of a species. A relatively common use of this occurs in the network literature where resistance is related to the number of other species extinctions (secondary extinctions) after the removal or addition of another species [e.g., (Dunne et al. 2004)].

Note: There is no real reason to expect persistence-based metrics to give the same answer as the change-based metrics above. In fact, one may expect the opposite. Let us imagine, for example, a resilient plant monoculture. After a small perturbation in its own density, the resilient monoculture returns rapidly because of its high growth rate. However, this same monoculture can also be enormously sensitive to an invading herbivore (e.g., its high growth rate correlates with its being highly edible) and so have a low resistance. As noted by Ives and Carpenter (2007), it behooves us to begin to consider these mutiple axes of stability and how nature struggles with different aspects of stability.

1.3 OF FOOD WEBS, STABILITY, AND FUNCTION

When considered at all scales, from a local patch to the entire biosphere, a food web governs the flux of energy and nutrients throughout our natural world. This flux and its fate ultimately drive a number of critical functions. Ecosystems recycle nutrients, decompose wastes, and produce primary and secondary biomass. All these major functions of an ecosystem ultimately service humans and frequently do so with such a consistency that people doubt ecosystems will ever stop serving them. Some of the services offered to humans are plant food, animal food, water purification, hospitable climate, detoxification, crop pollination, seed dispersal, wood, carbon storage, energy (e.g., water power, wood), and landscape stabilization. It is estimated that the services provided by natural ecosystems are worth more than 33 trillion U.S. dollars each year (Costanza et al., 1997). The actual number is debatable, but there is little doubt that ecosystems service humans in a way that we tend to overlook.

Stability, in the sense defined above, can be thought of as both a function and a property of an ecosystem. An imbalance of function can be costly to a society. If a fishery that depends on secondary production (e.g., the cod fishery of the North Atlantic) is lean for many years, this causes obvious problems for an economy. Similarly, outbreaks in pests and diseases incur great societal costs to agricultural crops and bodies of water (e.g., lost forestry or fisheries production). Both are examples of either a drastic reduction in a key species (e.g., cod) and/or a drastic increase in an unwanted organism (e.g., mountain pine beetle). They represent, at the least, a temporary loss in the consistency of the assemblage at both a local (e.g., 1-m^2) and a regional (e.g., ocean, forest stand) scale such that the system is tipped excessively one way or another. We now consider some examples of such documented instability and collapse that appear to be driven, at least to some extent, by human activity.

1.4   ECOLOGICAL INSTABILITY AND COLLAPSE

Rachel Carson's book *Silent Spring* was driven by a large-scale human action, the widespread application of DDT (Carson, 1962). One of the odd things about human impact is that it can occur at such enormous spatial and temporal scales. As DDT made bird eggs less viable, this global human impact directly knocked out a group of highly visible species. There was no refuge in space or time from the DDT, so the influence had direct consequences for birds and many other vulnerable species. Many other influences of DDT, though, were likely less direct and less obvious. I am not aware of anyone considering how the reduction in birds and other vulnerable organisms changed the rest of the ecosystem. One can imagine a huge array of potential indirect cascades throughout terrestrial and aquatic food webs. Perhaps nothing so dramatic occurred. It is possible that other organisms competing with birds rose slightly in density with the simultaneous reduction in birds. If so, nature may have had a way of balancing itself in the face of such a massive human perturbation, a way we just did not empirically detect. Unfortunately, we do not know what happened. We simply tend to hope that such human-induced perturbations find a way of working themselves out in a manner that does not influence our way of living.

Perhaps ecosystems are so robust that they almost always work themselves out without strongly influencing us, or perhaps we humans are just blissfully unaware that we are eroding critical components of an ecological system—the very attributes that allow the system to right itself in the face of our massive earth experiments. Even one counterexample forces us to recognize that instability and collapse are a real possibility. There are, in fact, numerous suggestions that we are tipping the scales of nature's balance. Below, I cover just a few of the emerging examples to highlight the fact that at least some human actions seem to be intricately linked to a loss in stability that results in the collapse of whole ecosystems.

1.4.1 NUTRIENTS ON THE LANDSCAPE

One commonly observed driver of ecosystem collapse is the large-scale anthropogenic movement of nutrients on the landscape. In a series of influential papers, empirical ecologists [reviewed in Polis and Holt (1992)] made the point clear that ecological systems are never isolated. Frequently, for example, nutrients are shunted around the landscape from one ecosystem to another, and in many cases this shunting of nutrients can play a large role in the dynamics of an ecosystem (Polis and Holt, 1992). Large-scale human impact and examples of instability and imbalance often fall within this subsidy framework. A classic example of such ecological instability and collapse driven by a human subsidy was brought to the attention of the ecological world by Bob Jeffries and his colleagues (Jefferies et al., 1994, 2004). Their long-term research on the Hudson Bay snow geese populations allowed them to closely monitor explosive population growth in snow geese. This explosive growth in herbivores has led to destructive overgrazing, an enormous loss of diversity, and a completely altered ecosystem (figure 1.1).

The cause of this unbridled growth in snow geese appears to be driven by a human subsidy that arises thousands of miles south of the impacted Hudson Bay lowlands. Population densities of snow geese are strongly correlated with the intensive increase in the application of nitrogen to cereal crops in the southern states (Jefferies et al., 2004). Nutrient-rich fields of cereal crops allow the snow geese to thrive before they leave in flocks for northern nesting grounds.

FIGURE 1.1. The destructive power of herbivores as shown by Jefferies et al. (2004). The top left photo is a relatively pristine ecosystem. All the others show the destructive power of herbivory. The bottom left photo contrasts a rich plant community (exclosure) against the barren landscape after herbivory. This is an example of what Don Strong has referred to as "runaway consumption' (Strong, 1992). Photo courtesy of Bob Jefferies.

The luxurious southern bounty of cereal crops manifests as healthy snow geese offspring that will ultimately return south to thrive on the rich crops before moving north to breed and consume the marsh. The result of this boom in geese has been devastating for the Hudson Bay lowlands. Lush fields of sedges have been replaced by salty nutrient-poor mud flats (figure 1.1). Geese are destructive feeders, removing even the roots of plants in an activity called *grubbing*. With the loss of plants and their roots, the soil is eventually destabilized, and so water running over the landscape carries off rich topsoil and nutrients in the process. The loss of plants in turn also exposes the soil to the sun. Under such conditions the salinity of the soil, is substantially increased and the ecosystem becomes inhospitable to most plant life.

In sum, runaway herbivory has led to ecosystem level feedbacks that have reduced this once-rich system to a barren landscape. Sadly, it appears that reversing the ill effects of this human subsidy will not be trivial. This research not only highlights how a human-driven subsidy has led to massive habitat destruction but is also a salient example of the enormous scale that ecological systems operate over. The world is indeed intimately connected.

Nutrient runoff also influences recipient aquatic ecosystems. Lake Erie and the Gulf of Mexico are two of many examples where terrestrial nutrient runoff fuels algal growth that cannot be rapidly assimilated by the pelagic component of the aquatic ecosystem (Dodds, 2006; Wilhelm et al., 2006). These unconsumed algae end up in the sediment, where they foster an explosive growth of bacteria. The bacterial boom in turn causes an oxygen deficit in the depths of aquatic ecosystems that can extend over enormous spatial scales in large bodies of water. In the face of such harsh conditions, many bottom-dwelling (benthic) organisms die and consequently reduce the flux of biomass up to higher-order predators. Thus, the direct reduction in benthic biomass and production likely cascades through these aquatic food webs, although the full extent of this cascade is unknown. These areas have become appropriately referred to as *dead zones* and are not trivial in size. For example, the Mississippi River drains enough nutrients into the Gulf of Mexico, to produce a dead zone of over $100 \, km^2$. Dead zones are examples of runaway microbial activity, a seemingly common phenomenon in collapsed ecosystems (both natural and laboratory-created ecosystems, as mentioned earlier).

Intact seagrass beds are wonderfully diverse and productive coastal ecosystems, but natural variation in human densities around different beds has repeatedly created gradients in species diversity (figure 1.2a). Alex Tewfik, a student and colleague of mine, found that as human density increased, these ecosystems lost top predators, detritus, specialist consumers, and edible seagrass (Tewfik et al., 2007a; Tewfik et al., 2007b). At high enough human densities, the systems are reduced to a homogenized habitat dominated by a single relatively inedible seagrass and an explosion of sea urchins (figure 1.2a–d). The reason for this changing structure is not fully known, but early findings suggest that generalist urchins consume increased algal production driven by the increased loading of nutrients in the water around heavily populated coastal areas. The human-induced nutrient subsidy, with the simultaneous loss of top-level predators through culling, promotes an elevated density of urchins, which drives runaway herbivory and extreme habitat homogenization.

An idea lurking behind the above stories is that certain organisms are poised to take advantage of human subsidies. When they do, they gain an excessive advantage and explode in biomass. This imbalance drives a cascade of consequences, which usually means a great loss of diversity and homogenization of the habitat. Further, in all the above cases, there are undoubtedly enormous human costs that accompany these collapses.

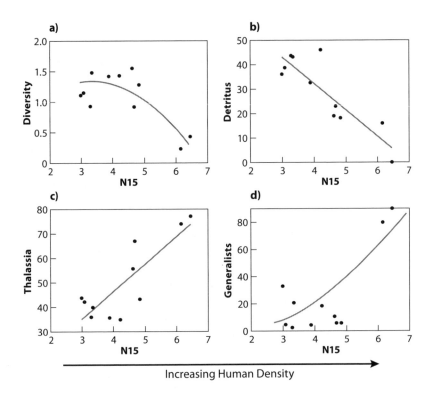

FIGURE 1.2. Some changing attributes of seagrass food webs across a gradient in the nitrogen isotope N-15 which correlates with human density (Tewfik et al., 2007a). (a) Data show a sudden jump to low diversity at high N-15 levels. (b) Detritus declines steadily with increases in N-15. (c) The relatively inedible seagrass *Thalassia* increases steadily with N-15. Not shown, the more edible *Syringodium* declines. (d) Generalist consumers show a sudden leap to extremely high densities at high N-15, mostly urchins at the endpoint. Figure modified from Tewfik et al. (2007a).

1.4.2 HARVESTING, FRAGMENTATION, AND RUNAWAY HERBIVORY

While the previous section outlined the bottom-up influences of nutrients on whole food webs, there is also abundant evidence that humans impact the top of the food web. Harvesting, for example, tends to be focused on the higher-order predators in both terrestrial and aquatic landscapes (Pauly et al., 1998; Jackson, 2001; Terborgh et al., 2001). Further, because these higher trophic level organisms are frequently large and mobile relative to lower trophic level organisms, they are also often the most negatively influenced by habitat fragmentation.

In an excellent review of human impact on coastal ecosystems, Jeremy Jackson and colleagues documented in detail the historical removal of large consumers by humans in coastal marine ecosystems and the subsequent delayed collapse of these food webs (Jackson, 2001). They argued using historical data that ecological extinction caused by overfishing precedes that caused by all other human disturbances in coastal ecosystems. Further, there is often a significant lag in time before ecosystems collapse, simply because unfished species take over the role of lost species until they too are negatively influenced by exploitation or disease. It appears as though changes induced by humans in coastal ecosystems have tended to drive a preponderance of microbial production and dominance (Jackson, 2001), much like the fate of most microcosm studies and the aquatic dead zones discussed above. Jackson (2001) referred to this as "microbilization" of the coastal ocean (Wilhelm et al., 2006). Perhaps, as suggested earlier, this microbial dominance is the signature of a complete collapse of a once-diverse ecosystem.

A recent well-explored example of the effects of such species removal has been uncovered in Yellowstone National Park. In Yellowstone, wolves, one of the major historical apex predators in the ecosystem, were completely eliminated by hunting by the 1930s. Fairly recently, wolves have been reintroduced into the park. The history of the park with and without wolves has been followed closely by a number of ecologists [e.g., Ripple et al. (2001)]. Once they were gone, the loss of wolves cascaded through the ecosystem, influencing both density and diversity in the park. Where the system consisted of a broad range of herbivores, elk grew in density to dominate herbivores in the absence of wolves. Further, the elk, no longer fearing predation, began to move into riparian areas they normally would not have visited in the presence of wolves (Ripple et al., 2001). The increased densities of elk and their accompanying consumption impact on riparian vegetation in turn has greatly altered the riparian zone. The Yellowstone story is reminiscent of Aldo Leopold's essay, "Thinking Like a Mountain", which predicted the runaway consumption impacts of herbivores and great losses in diversity resulting from the removal of wolves. Yellowstone has recently reintroduced wolves, and this reintroduction has proven relatively successful. Elk, now fearing predation near the more open riparian zone, have adjusted by moving higher in the landscape, leaving riparian vegetation to flourish (Ripple et al., 2001).

In another example, Terborgh and colleagues (Terborgh et al., 2001) took advantage of a hydroelectric dam that flooded a large terrestrial area in the tropics. The dam flooded the landscape such that the highland areas became islands surrounded by a sea of water. Top predators were eliminated from most islands (especially the small islands) but not the mainland, allowing comparison of the influence predators had on overall ecosystem structure. Again, herbivore abundance increased enormously in the absence of predators; predator-free islands had herbivore densities several orders of magnitude (10–100 times) higher than those in nearby mainland areas with top predators (Terborgh et al., 2001). This heightened herbivore density, not surprisingly, had dramatic implications for the density of understory seedlings and saplings. Again, top predators appeared to be playing a major role in promoting the persistence of a given assemblage of species. The collapse of diverse pristine assemblages seems to be relatively common when top predators are removed. Later in this book, I will return to some of these cases to suggest that this impact may be dependent on ecosystem size (McCann et al., 2005).

1.4.3   WATER IMBALANCE AND ECOSYSTEM COLLAPSE

Humans also frequently affect the balance of water on the landscape with dramatic food web consequences. The Aral Sea in Central Asia, at one time one of the largest inland seas on the planet, has seen a change of horrific proportions (Smith, 1994; Assessment, 2005). In an attempt to increase cotton production in the region, longstanding waterways were rerouted to croplands. This change in water use reduced the Aral Sea to a fraction of its original size. This dramatic change led to an enormous loss of native plants in the area surrounding the sea. With the loss of plants, the soil dried out and soil erosion increased, so that rain moving over the landscape took with it critical nutrients. Again, as in the snow geese example, the dry, sun-baked soil then increased in salinity. There are even large-scale climatic feedbacks that contribute to such "desertification" (Assessment, 2005). Windswept, barren areas increase the heat and dryness of the area and so reduce precipitation. This type of positive feedback is common to what ecologists now refer to as "alternative states" (Scheffer et al., 2001; Folke et al., 2004). Essentially, systems are changed enough that they cross a tipping point where the new system has positive feedbacks (e.g., loss of soil nutrients, inhospitable saline soil, and a dryer climate) that suddenly preserve it in its altered state. In the Aral Sea example, there are even arguments that in this region disease has become rampant and entire fisheries have been lost (increased water salinity likely contributes to this)! The region has been dealt enormous economic and social blows.

Restoration of a system that does not want to change easily obviously becomes an expensive endeavor. Examples of desertification (sometimes from increased herbivory) abound (Srivastava and Jefferies, 1996), and other examples of ecosystems switching into very different, often economically unfortunate states are emerging (Scheffer et al., 2001; Folke et al., 2004; Assessment, 2005).

1.4.4   HUMAN ECOSYSTEMS AND THE PARADOX OF ENRICHMENT

We have to this point discussed the impact of humans on natural ecosystems. There is also the case of human-made ecosystems that are frequently relatively spatially unstructured compared to their more natural counterparts (e.g., monocultures of plant crops, aggregations of fish for food, and monocultures of trees for lumber). In a sense, these homogenized ecosystems represent an extreme departure from pristine ecosystems where spatial and temporal variations seem to be the rule (Levin, 1999). As such, human-made ecosystems can be seen as an experiment in what happens when systems become strongly homogenized. While maximizing the area of a wanted biological product, humans have a long and costly history of defending these ecosystems against nature clawing back.

Agricultural and silvicultural ecosystems, for example, are frequently subjected to enormous outbreaks of pest species (Cappuccino et al., 1998; Logan et al., 2003). These pest species are fueled dynamically by the excess of resources concentrated in space (i.e., a nutrient-subsidized crop). Their populations explode locally and then spreads across the homogenized landscape like a wildfire. Fish farms have had similar issues, with the recent documentation of increased parasitic loads, disease, and contamination (deBruyn et al., 2006). In some cases, penned fish with their heightened disease burden have been shown to transmit contagion to nearby wild fish. There is a theoretical idea in ecology called the paradox of enrichment. I will discuss it in detail in this book, and it will become a foundation for the ideas laid out here. This idea is presented in many ways, but in its most qualitative presentation it suggests that predator-prey interactions in a well-mixed productive setting ought to readily drive boom-and-bust dynamics (i.e., an unstable situation by persistence-based metrics). In a sense then, boom dynamics are arguably a manifestation of the first half of this theory (i.e., that predators, pests, or diseases explode in number under such conditions). However, the second half of the theory—that resource populations are ultimately suppressed—is not always fully realized because humans go to great cost to alleviate the suppressive pressure of such pests.

1.5   A THEORY FOR FOOD WEBS

Here I have briefly laid out examples where humans have appeared to ignite collapses in ecological systems. I have also occasionally emphasized that this imbalance probably comes at a significant cost. Real ecosystems are indeed collapsing on the landscape. It therefore remains an important task to understand these complex natural entities in relation to their modified counterparts (i.e., urban ecosystems).

TABLE 1.1. Some Literature Examples of Major Ecosystem Impacts on an Ecosystem Following Some Perturbation(s)

Ecosystem	Perturbation	Impacts	Citation
Tropical islands	Hydroelectric dam (loss of top predators)	Runaway herbivory	Terborgh et al. (2001)
Yellowstone National Park	Loss of top predators (wolves)	Runaway herbivory	Ripple et al. (2001)
Marine ecosystems	Fishing (loss of top predators)	Unknown	Myers and Worm (2003)
Coral reefs and Estuaries	Fishing and eutrophification	Microbes dominate	Jackson et al. (2001)
Kelp forests	Loss of top predators	Runaway herbivory	Tegner and Dayton (2000)
Seagrass beds	Loss of predators and eutrophification	Runaway consumption	Tewfik et al. (2007a)

The examples presented above argue that human changes often reduce diversity, homogenize, and fragment the landscape. This creates wholesale rerouting of energy and nutrient fluxes throughout food webs such that a once-minor species can become a dominate biomass on the landscape. These changes alter feedbacks at local to global scales, feedbacks that are not easily predicted. At the very least, we are experiencing a period of great loss of natural beauty, and at the most, we are destroying the underlying life support we take for granted one ecosystem at a time.

Having laid out arguments that understanding the stability of ecosystems is a fundamental societal need, the rest of this book endeavors to understand how complex ecological systems work. I believe that our increased ability to understand these complex entities will ultimately form the conceptual infrastructure we need to deal with this rapidly changing world that is currently undergoing major habitat fragmentation, climatic change, and significant biodiversity loss.

This book proceeds by first presenting an introduction to dynamical systems theory. Here, my intention is to simply allow readers entry into this seemingly formidable area of math. As a result, I do not go into technical detail (although I will give references for the curious reader) but rather give the reader an intuitive walk through this fascinating area of math. The basic ideas behind the math are often quite simple, and I believe they are accessible to all readers. The ideas discussed here are then used throughout the remainder of the book in order to develop a theory about food webs.

In the second section of the book I examine the results from single-species models to small subsystem models (i.e., two-, three-, and four-species models, now commonly referred to as *modules*). In this part of the book I emphasize the role of interaction strength in the dynamics and develop some energetically driven principles that will form a backbone for more complex food web theory. The final section will expand upon this simple theory by extending the principles from modular theory to fit within a whole ecosystem context and all its accompanying complexities. Wherever possible I relate empirical and experimental results to the theory. Laid out in this way the book synthesizes results from single population models (including age structure) to whole ecosystem models that include nutrient recycling. I argue that there is a surprisingly coherent theory about food webs and ecosystems that sets the framework for understanding how perturbations (such as human impact) ought to influence the sustainability of ecosystems.

CHAPTER TWO

A Primer for Dynamical Systems

This chapter gives a brief introduction to the branch of mathematics known as *dynamical systems*. My hope is that even the less mathematically oriented reader will pass through this section in order to facilitate understanding of the theory developed in the rest of the book. This chapter is in no way a comprehensive review of dynamical systems. Rather, it is meant to establish familiar terminology as well as to make the reader comfortable with the broad concepts and approaches behind dynamical systems theory. I will argue that most of the mathematics and their graphical representation can be viewed through the eyes of an experimentalist. That is, things like bifurcation plots merely succinctly summarize the results of an experimental manipulation by modifying a key biological parameter. As a result of this emphasis, mathematical detail is often sidestepped (many excellent books already exist for this purpose), and instead I focus on some of the main conceptual ideas behind dynamical systems theory. I think these concepts are understandable to readers of all ilk, and sometimes technical details overwhelm people to such an extent that they miss overarching, and relatively simple, unifying concepts. The chapter ends with an empirical example from ecology in order to illustrate the ideas reviewed here and point out that the mathematics of dynamical systems underlies the dynamics of real ecological systems.

2.1 QUALITATIVE APPROACHES TO COMPLEX PROBLEMS

Everything living is a dynamical entity. Organisms respire, reproduce, disperse, and decompose, all processes that cause change in time and/or space. It is in this sense that the study of biological systems is evidently deeply dynamical in character. Fortunately for such a science, there is a well-developed branch of mathematics that deals specifically with dynamical systems, not surprisingly called *dynamical systems theory*, and this formalism is central to understanding the theory of biological systems. In this chapter I will give a brief background of dynamical systems and emphasize along the way the qualitative nature of this theory. While much has been made of the bounty of complex dynamics that can be produced by a mathematical system of equations, I will concentrate more on the notion that the actual types of qualitative solutions are enormously limited. Further, I will present these ideas from the perspective of an experimentalist, as a lot of dynamical techniques (e.g., bifurcation plots) are, in a sense, a summary of many independent experiments.

In nature, when we study a complex system, it is almost always immediately essential that we reduce the dimensions of the problem to something tractable. When we can do this with limited loss of information, we have made considerable progress. At a level, this is the job of a scientist: to tell a mostly true story of a complex natural system using the barest essentials or fewest dimensions. This ability to reduce a complex problem was one of the legacies of the great mathematician Henri Poincaré. He was one of the forefathers of dynamical systems theory, a fascinating branch of mathematics that seeks to delineate and understand the behavior of dynamical systems (i.e., ordinary differential equations, partial differential equations). Until Poincaré, mathematicians had had much success with linear dynamical systems by effectively solving each problem independently, but a similar approach to nonlinear problems seemed hopeless. Nonlinear problems brought with them endless permutations that confounded the classic approach.

Faced with this vexing problem, Poincaré decided to change the approach by introducing a more qualitative analysis that, grossly stated, categorized dynamics into groups based on topological principles. Here, he changed the focus away from exact solutions toward more qualitative questions like Does the long-term behavior of the system oscillate? [see Barrow-Green (1997) for an interesting historical account of Poincaré's contributions]. The approach forged a path to a general understanding of dynamical systems. It is worthwhile considering this successful tactic as a methodology for such a complex science as ecology.

The theme of reducing dimensions according to underlying similarities, or biological constraints, will permeate this book. The approach set out here is therefore often qualitative in that the theory often predicts that one food web structure is more or less stable than another. I must emphasize that this does not mean it is not testable or amenable to empirical analysis. I will frequently discuss how the qualitative theory of food webs presented in this book predicts testable patterns. Where possible, I will bring in data to make explicit connections between qualitative theory and empirical data.

In order to garner an intuitive understanding of the theory that follows, I now present a primer, largely graphical in nature, on dynamical systems and bifurcations. *Bifurcations* are simply transitions from one qualitative dynamical outcome (e.g., steady-state dynamics) to another qualitative outcome (e.g., oscillations). For more detailed definitions and explanations, please refer to the excellent books available on this topic [e.g., Guckenheimer and Holmes (1983); Wiggins (1990)].[1] Where possible, I will employ familiar ecological models to highlight the theory of dynamical systems.

2.2 DYNAMICAL SYSTEMS

2.2.1 THE PHASE SPACE AND TIME SERIES

A common graphical technique in consumer-resource interaction theory is a *phase space* diagram, which is accompanied by a consumer and a resource *isocline*. This graphical approach at first seems complicated, largely because the dynamics are implicit in the graph. As a result there is no dimension for time respresented on the graph. In the same sense that scientists employ graphs to readily convey a set of ideas, mathematicians employ phase space diagrams to convey an understanding of the dynamics of a set of equations. In order to motivate this primer on dynamical systems and the use of phase space diagrams, let us start by considering the basic Rosenzweig-MacArthur (R-M) consumer-resource (C-R) model that assumes logistic resource growth and a type II function response by the consumer (i.e., consumption rate saturates with resource density):

$$
\begin{cases}
\dfrac{dR}{dt} = rR(1 - R/K) - \dfrac{aCR}{R + R_o}, \\[2mm]
\dfrac{dC}{dt} = e\dfrac{aCR}{R + R_o} - mC,
\end{cases}
\tag{2.1}
$$

where r is the per capita rate of increase in resource, R, K is the carrying capacity of the resource, a is the attack rate of the consumer, C, on R, e is the conversion efficiency of R into C, R_o is the half-saturation density, and m is the mortality rate of the consumer, C. This is arguably the most common consumer-resource model. See Hastings et al. (1997) for a detailed account of the formulation and biological assumptions behind this model.

Figure 2.1a shows an example of the phase space and the accompanying isoclines of this system for a specific set of parameter values. The phase space in this case is a two-dimensional graph of all possible resource densities, R, and

---

[1] There is a wonderfully illustrated series of books by Ralph Abraham and Chris Shaw, called *Dynamics: The Geometry of Behavior*, that allows a general audience a visual entry point into the theory of dynamical systems by using almost nothing but colorful, clear illustrations (Abraham and Shaw, 1982).

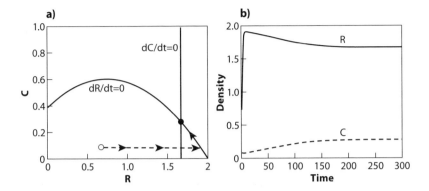

FIGURE 2.1. Dynamical representations of model equation (2.1). (a) The phase space for consumer density, C, and resource density, R. The isoclines are also represented. Given an initial value (open circle), the dynamics follow the trajectory depicted in the phase space. Time is implicit in the trajectory on the graph. (b) The actual time series sketched out by the resource density (dashed) and consumer density (solid) over the trajectory depicted in the phase space in (a) Parameter values: $r = K = 1$; $a = 1.30$; $b = e = m = 0.50$.

consumer densities, C. The resource isocline traces out all the combinations of C and R that balance the resource growth with respect to time (i.e., where $dR/dt = 0$). Thus, the R isocline in figure 2.1a shows combinations of C and R where the resource dynamics do not change. Note that the density of C is free to change for combinations of C and R that lie on the R isocline, as this isocline says nothing about consumer dynamics. Consumer dynamics are instead held in check by combinations of C and R on the C-isocline curve in figure 2.1a (i.e., where $dC/dt = 0$). Clearly, wherever the two isoclines intersect, we have identified a combination of C and R where both populations simultaneously do not change (dR/dt and $dC/dt = 0$). This is an example of what scientists refer to as an *equilibrium* (see solid circle in figure 2.1a).

Now let us see how the model system (2.1) behaves dynamically for a given set of parameter values. To run this model we must have an *initial value* of R and of C, which then change in time deterministically according to equation (2.1). This is really no different than an experiment—we must have a starting point. The open circle in figure 2.1a identifies the initial value, and the trajectory indicates how this system, given the parameters, changes in terms of R and C with time. Time is not explicit in this graph, but we know that it increases along the trajectory. To see this, we can also choose to represent this solution as a time series where we follow either the consumer or the resource densities as a function of time. Figure 2.1b shows the corresponding time series. You understand this subtle point if you can make sense of the

---

BOX 2.1

SOME IMPORTANT DEFINITIONS

State variables: The variables that are defined dynamically in the equations. In equation (2.1), these are R and C.

Parameters: The variables that are independent of the equations. In equation (2.1) these are a, r, e, R_o, and m.

Phase space: A space defined by the state variables. (e.g., C and R in equation (2.1)).

Isoclines: The combination of values of the state variables (C and R here) that yield zero solutions for a given dynamical equation (e.g., those values of C and R that give $dR/dt = 0$ yield the R isocline).

Initial value: The choice of values of all state variables at time $t = 0$.

Equilibrium steady state: A point in the phase space where all rates of change are zero (i.e., isoclines intersect) such that if the trajectory landed *exactly* on that point, it would remain there forever. An equilibrium can be an attractor, neutral, a repellor, or saddle (defined below). Represented in phase space by example figure 2.2a.

Nonequilibrium steady state: A set of points in the phase space where the average rate of change over the set of points is zero such that if the trajectory landed *exactly* on one of these points, it would remain within this set forever. A nonequilibrium steady state can be an attractor, neutral, a repellor, or a saddle (defined below). We define two different manifestations of nonequilbrium states:

  (i) **Periodic oscillation**: A trajectory in phase space that repeats itself (e.g., consumer-resource limit cycle). A periodic oscillation can be an attractor, neutral, a repellor or a saddle. Represented in phase space by example figure 2.2b.

  (ii) **Complex oscillations**: A trajectory in phase space that does not repeat itself but has almost repeatable structure (e.g., chaos, quasiperiodic dynamics). Such nonperiodic dynamics can also be attractors, neutral, repellors, or saddles. They are represented in phase space by figure 2.2c.

Transient: The portion of the time series before the dynamics settle down on the attractor. In figure 2.1b the transient portion is all the dynamics prior to the point where the solution reaches the equilibrium and then flatlines (i.e., stays the same for all time).

Attractor: A point (equilibrium steady state) or set of points (e.g., a nonequilibrium steady state) in the phase space that a dynamical system, once given an initial value, ends up at and then remains. Different initial values may end up at different attractors in a given dynamical system.

Repellor: Any equilibrium point or periodic or complex oscillation can be a repellor if all nearby points in the phase space move away from the steady state. This is an unstable steady state.

Saddle: Any equilibrium point or periodic or complex oscillation can be a saddle if nearby points attract in some dimensions but repel in other dimensions of the phase space. This is an unstable steady state like a repellor because all trajectories near the equilibrium ultimately depart.

translation between the representation of the dynamics in figure 2.1a and b. The rest of this chapter will focus on representing solutions and dynamics in the phase space. Box 2.1 reviews the mathematical definitions and terms used in this book.

2.2.2   EQUILIBRIUM AND NONEQUILIBRIUM STEADY STATES

Figure 2.1b shows a time series that ends up at an attracting dynamic equilibrium; the solution is eventually flatlined such that it no longer changes. This is an example of a dynamic equilbrium because there is still a flux, but the "flux in" is matched exactly by the "flux out"—it is in perfect balance. Here, we realize that the equilibrium is a single point in a phase space representation as shown in both figures 2.1a and 2.2a. Nonequilibrium steady states also are balanced in time but slightly differently. Rather than being a single point, the steady-state solution is a set of points where the average rate of growth over the entire set is zero. The phase space respresentations of two well-known types of nonequilibrium steady states are shown in figure 2.2b (limit cycle) and c (chaotic attractor). All nonequilibrium steady states produce fluctuating time series, which have a long-term average population growth rate that is effectively zero. The time series for the limit cycle, for example, is the well-known consumer-resource or predator-prey cycle of introductory population ecology books that is easily produced by the Rosenzweig-MacArthur model [e.g., see Hastings et al. (1997)].

Once we define a steady state, say an equilibrium, it is natural to think about what happens to the dynamics of the system if we perturb the densities off that equilibrium steady state. Mathematicians have recognized that this question is quite answerable if we make the perturbation off the equilibrium very small. It turns out that there are only three qualitatively different answers to this problem: (1) the densities return to the equilibrium or the equilibrium acts as a *stable attractor* (figure 2.3a); (2) the densities do not move anywhere or the equilibrium is *neutral* (figure 2.3b); and (3) the densities move away from the equilibrium, and so we have an *unstable equilibrium steady state* (figure 2.3c). Qualitatively speaking, there are no other options.

The case in which the equilibrium is unstable requires further comment. Often an equilibrium will not repel in all directions. As an example, figure 2.4 shows two different cases where an unstable equilibrium steady state repels in all directions (figure 2.4a; equilibrium is a repellor) and a case where the unstable equilibrium steady state actually repels in only one direction (figure 2.4b). This latter case is often referred to as a *saddle node*. Dynamically, saddles are important, as they separate different basins of attraction. That is, some trajectories depart to one attractor, while trajectories on the other side of the saddle depart to a different attractor (i.e., an alternative state). The trajectory separating these basins shown in figure 2.4b is referred to as a *separatrix* because it separates two different basins of attraction. Clearly, there exist nonequilibrium repellors and saddles, and for simplicity I have shown just the equilibrium examples in figure 2.4.

These equilibrium steady-state results naturally extend to nonequilibrium steady states such that nonequilibrium dynamics (e.g., figure 2.3d) have the same properties as above. That is, the nonequilibrium steady state can attract (figure 2.3d), behave neutrally (figure 2.3e), or be unstable (e.g., figure 2.3f). While here I have identified limit cycles (figure 2.3d–f), other more complicated nonequilibrium steady states, like chaotic dynamics, exist with the same properties. As such, there are unstable chaotic steady states as well as chaotic attractors [see McCann and Yodzis (1994), Vandermeer and Yodzis (1999), and Schreiber (2001) for some ecologically motivated examples].

We have taken the set of all possible mathematical dynamical systems and reduced their dynamics to a rather limited set of dynamical possibilities. Specifically, we have identified two types of steady states (equilibrium and nonequilibrium) that both possess only three possible types of local dynamics (stable, neutral, or unstable). We are now in a position to discuss some of the qualitative techniques of dynamical systems theory.

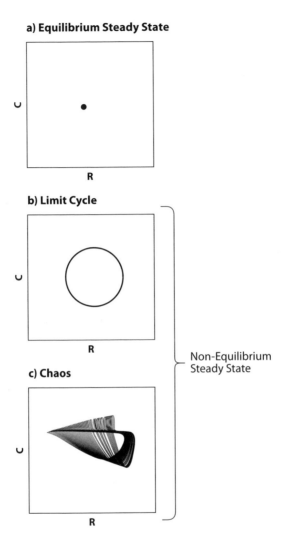

FIGURE 2.2. Dynamical representations in a phase space of the dynamics after a perturbation off three different steady states. (a) An equilibrium steady state. (b) A nonequilibrium steady state known as a limit cycle or oscillating solution. (c) A more complicated nonequilibrium steady state. This state is shown in two dimensions but is actually projected from a food chain model (P-C-R) onto the C-R phase space. For ordinary differential equation models this complicated trajectory can occur only in three or more dimensions because trajectories cannot cross in continuous models. Diagram of the chaotic attractor is from McCann and Yodzis (1994).

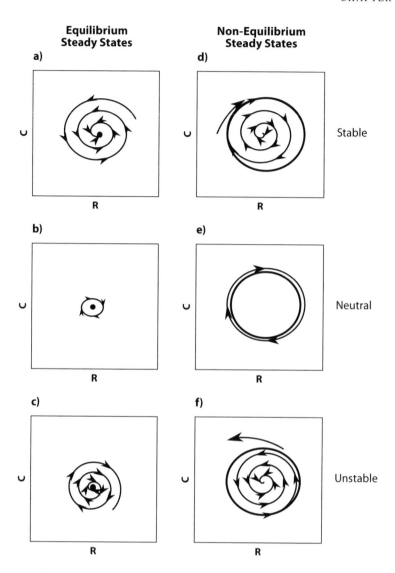

FIGURE 2.3. Dynamical representations in a phase space of three qualitatively different local dynamics of equilibrium [(a)–(c)] and nonequilibrium steady states [(d)–(f)]. (a) and (d) depict the locally attracting steady states, (b) and (e) the neutral case; and (c), (f) the unstable case. See figure 2.4 for the two types of unstable attractors.

Unstable Equilibria

a) Repellor   **b) Saddle**

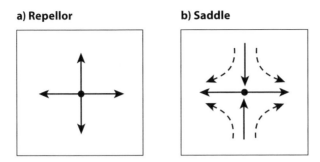

FIGURE 2.4. Dynamical representations in a phase space of (a) a repellor that repels in all directions and (b) a saddle node that repels in one dimension and attracts in another. Saddles act to separate different basins of attraction. Notice that trajectories shoot in two different directions, left and right. Ultimately these trajectories lead to different attracting states not shown.

2.2.3 THE QUALITATIVE ANALYSIS OF STEADY STATES: EIGENVALUES AND BEYOND

Nonlinear dynamical systems are most often not easily solved, and so one approach to understanding them is to restrict the analyses to very small perturbations off a steady state. We can readily consider such a small perturbation off an equilibrium steady state (R^*, C^*) by recasting the system such that we follow the distance from the steady state instead of following R and C [we convert system (2.1) into x and y, where $x = (R - R^*)$ and $y = (C - C^*)$]. All dynamics under this transformation are now easily interpreted in terms of whether the dynamics near the equilibrium are attracted toward or repelled away from the equilibrium steady state. Specifically, if x and y grow, the systems dynamics are repelling away from the equilibrium, while if x and y tend toward zero, the dynamics are being attracted to the equilibrium [i.e., $x = (R - R^*)$ approaches zero, so we are approaching R^* as time increases].

This restriction of the dynamics to values very close to the equilibrium is important. It is well known that any function, F, can be also represented by a Taylor series such that

$$\begin{cases} F = F(a) + \dfrac{F'(a)}{1!}(x - a) + \dfrac{F''(a)}{2!}(x - a)^2 + \cdots, \\ F = F(a) + \text{linear term} + \text{nonlinear terms (e.g., } x^2, x^3, \ldots), \end{cases} \tag{2.2}$$

where F' denotes the derivative of F evaluated at the point a, F'' denotes the second derivative of F evaluated at the point a, and so on. The take-home message from this result is that functions can be decomposed into a linear term, x, coupled to an enormous sum of nonlinear terms (x^2, x^3, etc.).

Knowing this, and knowing that we have restricted our analysis to values very close to R^*, we are in a position to make this problem simple. Since our linear term $x = R - R^*$ is small by definition, all the nonlinear terms that would accompany this in the appropriate Taylor series become negligibly small. Recall that a small number squared is an extremely small number, a small number cubed is even smaller, and so on. The analysis can therefore proceed by reducing our problem to just the linear terms; this part of the analysis is frequently called *linearization*.

This is basically what we are doing when calculating the eigenvalue of an equilibrium in a dynamical system. The eigenvalues are the linear representations of the nonlinear system near the equilibrium. The number of eigenvalues is generally equal to the number of equations. There is a fair amount of mathematical mechanics required to get to the eigenvalue, but it is really much like following a recipe [see Hastings et al. (1997) for a more thorough example of these calculations, as well as below for an explicit example]. The important point is that the system is linearized near the equilibrium (e.g., $dx/dt = \lambda x$) and ultimately yields dynamical solutions of the form

$$x(t) = x(0)e^{\lambda t}, \tag{2.3}$$

where λ is the eigenvalue determining whether the solution returns to the equilibrium or not. There will be as many equations like (2.3) describing the dynamics near the equilibrium as there are in the original model system (e.g., a C-R model has two equations of form (2.3). Note, once the system is greater than one dimension in continuous time (i.e., is following more than one population), the eigenvalue or linear term can be either a real number (e.g., $\lambda = 2$) or a complex number (e.g., $\lambda = 3 + 4i$). Complex numbers have both a real part (e.g., 3 in the preceding eigenvalue) and a complex part (e.g., $4i$ in the preceding eigenvalue)—this aspect of the eigenvalue is critical for determining the local dynamics.

The *real part* of any eigenvalue, λ, determines the local stability. In solution (2.3) a negative real part in λ means that as time, t, increases, the exponential terms decay to zero. In terms of the implications for the perturbation, $x(0)$, this means that the perturbation goes to zero. We therefore know that any local perturbation attracts back to the equilibrium when the real part is negative. Following similar logic, a positive real part moves the solution away from the equilibrium [the exponential necessarily grows in (2.3)]. We therefore know that any local perturbation repels away from the equilibrium when the real part is positive.

On the other hand, the *complex part* of the eigenvalue says nothing about whether the equilibrium is locally stable or not; however, the complex part determines if the solution behaves monotonically (when the complex part is zero) or fluctuates (when the complex part is nonzero). This is a fascinating aspect of eigenvalues. A complex eigenvalue effectively means that our solution has some spin on it (i.e., fluctuates around the equilibrium). Similar procedures exist for nonequilibrium steady states. One can just as readily perturb infinitesimally off a limit cycle and ask if it attracts or repels. The mathematics is trickier here and beyond the scope of this book, but the techniques are qualitatively similar (Wiggins, 1990).

I now illustrate some of the mechanics of local stability analysis for an equilibrium using the classic Lotka-Volterra model for an equilibrium steady state. Those not interested in the details can readily skip to the next section.

2.2.3.1   *An Illustrated Example: The Neutral Lotka-Volterra Model*

In order to illustrate the local stability analysis of an equilibrium, let us consider the classic Lotka-Volterra model. This model is close in spirit to the Rosenzweig-MacArthur consumer-resource model of equations (2.1) but has no regulatory term $(-rR^2/K)$ and uses a linear or type I functional response. This model produces the well-known result of neutral dynamics (later we will look at the phase space for this model as well). The equations are

$$\begin{cases} \dfrac{dR}{dt} = F_R = rR - aCR, \\[2mm] \dfrac{dC}{dt} = F_C = eaCR - mC, \end{cases} \tag{2.4}$$

where F_R and F_C are the resource and consumer functions and all parameters have been defined above. First, let us determine the interior equilibrium by setting equations (2.4) to zero and solving for R and C ($R^* = m/(ea)$ and $C^* = r/a$). Notice that substituting $R^* = m/(ea)$ and $C^* = r/a$ into equations (2.4) makes $dR/dt = 0$ and $dC/dt = 0$ simultaneously. Let us now consider the dynamics near this equilibrium steady state $[R^* = m/(ea); C^* = r/a]$. We can determine the linear terms for equations (2.4) by determining how the rate-of-change functions, F_R and F_C, change with respect to R and C at the equilibrium. The linearized version of equations (2.4) written in matrix form is

$$\begin{bmatrix} \dfrac{dx}{dt} \\[2mm] \dfrac{dy}{dt} \end{bmatrix} = \begin{bmatrix} \dfrac{\partial F_R}{\partial R} & \dfrac{\partial F_R}{\partial C} \\[3mm] \dfrac{\partial F_C}{\partial R} & \dfrac{\partial F_C}{\partial C} \end{bmatrix} \times \begin{bmatrix} x \\[2mm] y \end{bmatrix}, \tag{2.5}$$

where x describes the dynamics of $(R - R^*)$ and y describes the dynamics of $(C - C^*)$. Now we determine the specific partial derivatives from equations (2.4) and substitute them into (2.5), yielding

$$\begin{bmatrix} \dfrac{dx}{dt} \\ \dfrac{dy}{dt} \end{bmatrix} = \begin{bmatrix} r - aC & -aR \\ eaC & eaR - m \end{bmatrix} \times \begin{bmatrix} x \\ y \end{bmatrix}. \tag{2.6}$$

Finally, we substitute the equilibrium values, R^* and C^*, into equation (2.6):

$$\begin{bmatrix} \dfrac{dx}{dt} \\ \dfrac{dy}{dt} \end{bmatrix} = \begin{bmatrix} 0 & -m/e \\ er & 0 \end{bmatrix} \times \begin{bmatrix} x \\ y \end{bmatrix}. \tag{2.7}$$

At this point we have linearized the Lotka-Volterra equations (2.4) at the interior equilibrium such that we now have the simpler problem defined by (2.7). The matrix of partial differentials in equations (2.5)–(2.7) are frequently referred to as the *Jacobian matrix* or in ecology as the *community matrix* (because they describe the linear terms of species interactions). This linearized problem is relatively easily solved using cookbook techniques borrowed from linear algebra. Specifically, we diagonalize the matrix to put it in as "simple a form as possible." Fortunately, linear algebra gives us the tools to determine the values of the diagonal terms, called the eigenvalues, λ [see any introductory linear algebra text for details; e.g., Berberian (1992)]. One does this by first subtracting the eigenvalues from the diagonal terms in equations (2.7):

$$\begin{bmatrix} 0 - \lambda & -m/e \\ er & 0 - \lambda \end{bmatrix}, \tag{2.8}$$

from which we determine the characteristic polynomial (again, see any introductory linear algebra text for how characteristic polynomials are calculated). For the 2×2 matrix the characteristic polynomial is simply the diagonal terms multiplied together minus the off-diagonals multiplied together. Given this, we get the following characteristic polynomial:

$$\lambda^2 + mr = 0. \tag{2.9}$$

The above characteristic polynomial (a quadratic) has as its solution the purely imaginary eigenvalues $\lambda_1 = mr^{1/2}i$ and $\lambda_2 = -mr^{1/2}i$. The solution to the characteristic polynomial can be determined using any symbolic mathematics software or, for this relatively simple case, employing the well-known quadratic formula $\lambda_i = \dfrac{-b \pm \sqrt{b^2 - 4ac}}{2a}$.

Equation (2.7) therefore can now be cast in the following equivalent diagonalized form:

$$
\begin{bmatrix} \dfrac{d\epsilon_1}{dt} \\[2ex] \dfrac{d\epsilon_2}{dt} \end{bmatrix} = \begin{bmatrix} 0 + mr^{1/2}i & 0 \\ 0 & 0 - mr^{1/2}i \end{bmatrix} \times \begin{bmatrix} \epsilon_1 \\ \epsilon_2 \end{bmatrix}. \tag{2.10}
$$

Note that the diagonalization of the linearized matrix, which acted on x and y, transformed the problem yet again so that the coordinates are no longer in x and y but rather now in a final form we have notated as ϵ_i. These multiple transformations are a little mind-bending, but the real issue to concentrate on here is that the eigenvalues completely determine the dynamics near the equilibrium. The time series solutions to this local perturbation problem are therefore determined by

$$
\epsilon_1(t) = \epsilon_1(o)e^{[0+mr^{1/2}i]t},
$$
$$
\epsilon_2(t) = \epsilon_2(o)e^{[0-mr^{1/2}i]t}, \tag{2.11}
$$

where $\epsilon_i(0)$ is the small initial perturbation in this example.

Because the real parts of the eigenvalues are zero, or neutral, in equation (2.11), we know that all local perturbations neither approach zero (as they would if the real parts were negative, they would decay to zero) or repel (as they would if the real parts were positive, they would grow exponentially). In the terminology of dynamical systems, we say that the equilibrium is *neutrally stable*.

As discussed above, the complex portion of an eigenvalue allows us to determine whether solutions near the equilibrium fluctuate or not—a complex eigenvalue means the solution has spin on it. In the above case the complex part of λ implies that a small perturbation fluctuates around the equilibrium (C and R go up and down), while the zero real part of λ implies that it does so in a way that it never approaches or moves away from the equilibrium (figure 2.3b qualitatively depicts this dynamical case). We have just mathematically determined the famous result of neutral cycles often presented in introductory ecology classes.

2.2.4   FROM LOCAL TO GLOBAL

Up to this point we have concentrated on a specific equilibrium steady state when determining local stability. In most models there is more than one steady state that we will be concerned with. In the last section we changed the problem so that we considered only the dynamics near a given steady state. Therefore, we are still in a position where we do not understand the full global dynamical

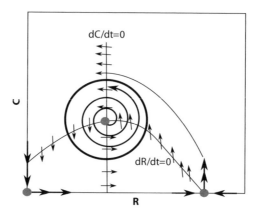

FIGURE 2.5. Dynamical representations in a phase space of a R-M consumer-resource phase space with three steady states [two locally unstable (repelling) equilibria and a locally stable (attracting) limit cycle]. Once all the local dynamics are determined, the global dynamical picture emerges because the global dynamics *must* make sense in light of the local dynamical phenomena. See the text for full a discussion.

behavior. One way to proceed is to categorize the dynamical behavior near all equilibria in a given dynamical system and then effectively piece together the global dynamical story.

Let us look at the Rosenzweig-MacArthur consumer-resource model [equation (2.1)] for an example of piecing local dynamics together into a global picture. Most consumer-resource models have an equilibrium at the origin ($R^* = 0$, $C^* = 0$, called the trivial equilbrium), an equilibrium at the resource's carrying capacity ($R^* = K$; $C^* = 0$), and finally an equilibrium with both positive consumer and resource densities (called the interior equilibrium; the equilibria are depicted with solid circles in figure 2.5).

In figure 2.5, the equilibria at the origin and at the carrying capacity are unstable such that a small positive perturbation grows. The population growth rate near zero resource densities and near zero consumer densities is positive, while the small addition of a consumer when the resource is at its carrying capacity is also positive. Figure 2.5 depicts these equilibria and their local stability properties. Now let us further consider the case where the interior equilibrium is unstable and there exists a stable nonequilibrium steady state or a limit cycle (figure 2.5).

At this point we have a lot of local information and given that we have all steady states and their local stability properties, we can piece together the global dynamical picture. Before we do this let us add the direction of the trajectories on all isoclines because the directions are easily determined. On the $dR/dt = 0$ isocline the trajectories in the R direction are by definition 0, but anywhere to the left of the $dC/dt = 0$ isocline we have negative rates of change for C and to the right of the $dC/dt = 0$ isocline we have positive rates of change for C. This means that we have vertical downward trajectories on the $dR/dt = 0$ isocline to the left, and upward trajectories to the right of the $dC/dt = 0$ isocline (figure 2.5). We can use similar logic to populate the $dC/dt = 0$ isocline with horizontal trajectories (figure 2.5).

It is apparent from inspection of this local information in figure 2.5 that the limit cycle is both locally and globally stable because any trajectory in the interior moves toward the limit cycle. Numerical simulations of consumer-resource models that fall into this case confirm this result. In a sense, then, the local stability results of the steady states are the pieces of the puzzle such that, when considered all together, the global dynamics emerge. This piecing of the local with the global requires that we know the local stability conditions on all equilibrium and nonequilbrium steady states absolutely, but sometimes this information is not so easy to obtain. Nonetheless, the point is that local dynamics sum to give us the global picture in some coherent fashion.

2.2.5   CONCEPTUAL EXPERIMENTS AND BIFURCATION PLOTS

Much recent theory employs bifurcation plots as a means of exploring the dynamics of mathematical models. While an obscure and formidable-sounding term to a nonmathematician, to a scientist a *bifurcation plot* can be thought of as a way of performing a conceptual experiment. As an example, one may look at how dynamics change as a function of the maximum attack rate by a con-sumer, a_{max}, in the R-M model [equation (2.1)]. To do this, one simply runs as many experiments (each experiment produces its own time series in this case) as one can. Figure 2.6a–c shows three different experiments, or time series runs, for three different values of a_{max}.

To obtain anything from these experiments we need to measure some aspect of the dynamics as a response variable. A common response variable in theory is the local maxima and local minima on the attractor. The local minima and maxima are chosen because they enable us to extract a lot of information from a solution in a simple manner. As an example, if the solution is an equilibrium (e.g., figure 2.6a and b), the local minima are the same as the local maxima. One point, the equilibrium value, then describes the dynamics well. This again is the mathematical trick of reducing a problem (many points in a solution) to a simple and informative representation (one point that defines the equilib-rium density). For nonequilibrium steady states, like a limit cycle (figure 2.6c), the local maxima and minima each have one point. Two points thus define

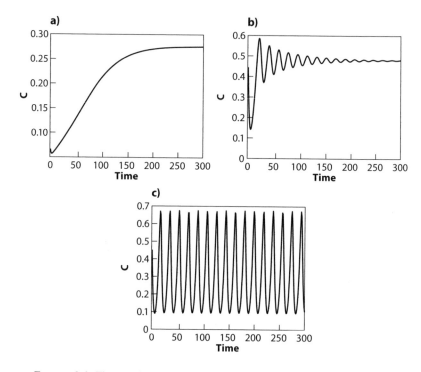

FIGURE 2.6. Time series of consumer density taken from the theoretical experiment described in the text for three different values of a_{max}. (a) Monotonic approach to equilibrium ($a_{max} = 1.20$). (b) Oscillatory decay to equilibrium ($a_{max} = 1.60$). This represents the onset of overshoot consumer-resource dynamics, an important distinction for many ecological problems. (c) An example of oscillatory dynamics that occur after the Hopf bifurcation ($a_{max} = 2.00$). Other parameter values: $r = 1.0$; $K = 2.0$; $e = R_o = m = 0.50$.

this nonequilibrium steady state well. On even more complex dynamics, say a chaotic time series, the local maxima and minima technique pulls out multiple points for local minima and local maxima. We will give an example of more complex dynamics later. Figure 2.7a shows the sum of many such experiments for the R-M model [equation (2.1)]. We see that a_{max} tends to destabilize the solution as it increases because the dynamics go from an equilibrium point to a limit cycle (at a high a_{max} there are two points for each a_{max} value indicating a limit cycle). The dashed lines indicate the "transition to oscillatory decay" and "oscillations or cycles" (as denoted by Hopf), respectively, in figure 2.7. Recall that the transition to oscillatory decay occurs when the eigenvalue becomes complex.

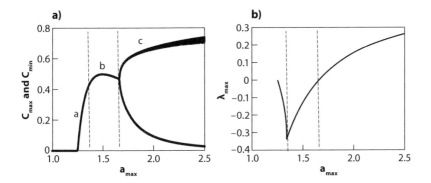

FIGURE 2.7. Theoretical experiment (bifurcation diagram) across the parameter carrying capacity, a_{max}; (a) Displays the local maxima and minima (i.e., where the time series curve down and up, respectively). Points identified as a, b, and c in bifurcation plots are plotted as time series in figure 2.6. Each data point in a bifurcation plot corresponds to a time series at the given parameter value, a_{max}. Two important dynamical changes, denoted by dashed lines, occur: (1) the transition to overshoot dynamics (a complex eigenvalue), and (2) the onset of cyclic dynamics (known as a Hopf biurcation). (b) The eigenvalue representation of this same bifurcation experiment. Note that the transition to complex eigenvalues and the Hopf bifurcation are the same as the representation in (a). Parameter values: $r = 1.0$; $K = 2.0$; $e = R_o = m = 0.50$.

Other response variables can be used, such as the eigenvalue, λ_{max}, of an interior equilibrium. Either way, the bifurcation plot shows important dynamical information (i.e., maxima, local stability) and along the way identifies important transitions between qualitatively different dynamical regimes. For example, in figure 2.7a and b, the dynamics switch from stable equilibrium dynamics to oscillations at the point denoted as the *Hopf bifurcation*. This materializes as a switch to two points in the maxima-minima plot (figure 2.7a), while this same transition is denoted in figure 2.7b at the switch from a negative eigenvalue to a positive eigenvalue [see Wiggins (1990) for more information on Hopf bifurcations and their conditions].

Further, I have also identified the parameter values that yielded the time series in figure 2.6a–c to aid in interpretation of the bifurcation plot. The purely real eigenvalue (a in figure 2.7a) yielded the monotonic dynamical time series example shown in figure 2.6a, b in figure 2.7a denotes the a_{max} value that yielded the time series for a complex eigenvalue with a negative real part (the oscillating solution decays to the equilibrium as expected in figure 2.6b), while c in figure 2.7a denotes the cyclic time series in figure 2.6c (a complex eigenvalue with a positive real part). The dynamics go to a limit cycle, reminiscent of the phase space discussed in the last section.

In what follows, we will find that the transition from monotonic to oscillatory decay often has powerful implications for understanding when a biological parameter is stabilizing or not. As an example, in figure 2.7b, we see that when the eigenvalue is purely real, the solutions become increasingly more stable with an increasing attack rate (i.e., the maximum eigenvalue gets more negative, which means it shoots back to the equilibrium quicker); however, as soon as the dynamics transition to complex dynamics, the attack rate destabilizes the dynamics because the eigenvalues get increasingly more positive with increased attack rates. As mentioned above, this transition is the first dashed line in figure 2.7a and b. In essence, the transition to complex eigenvalues implies that the consumer-resource interaction now displays overshoot dynamics (C rises so high it suppresses R, and a suppressed R means C collapses, allowing R to rebound, and so on).

In summary, to obtain a bifurcation plot we proceed in the following way, analagous to an experiment:

1. Vary the independent parameter (e.g., attack rate, a_{max}). Numerically, we can often perform many experiments because computational power is so fast. Figure 2.6 shows some examples of specific experiments for three different values of the attack rate, a_{max};
2. Measure some informative aspect of the response variable (here, the consumer and resource dynamics) from each experiment (e.g., local maxima and minima). In this case, like the experimentalist, we allow each experiment to run through its transient stage and then record data once it essentially lies on the attracting steady state;
3. Plot the response variable versus the independent variable (i.e., the bifurcation plot, figure 2.7a). The three time series depicted in figure 2.6, for example, are denoted in figure 2.7a by a–c.

As with any technique, application can become more complicated. There are bifurcation experiments, for example, that vary more then one independent variable and track when the equilibrium is stable or unstable (e.g., a curve describing where the Hopf bifuraction occurs in parameter space). Other fascinating bifurcations can also be documented (e.g., saddle node, transcritical, cyclic Hopf). These more complicated experiments are beyond the scope of this book but can be powerful for fully understanding the entire dynamical repertoire of a model system [see Guckenheimer and Holmes (1983), Wiggins (1990)].

2.2.6   The Relationship between Geometry and Dynamics

Theory often explores mathematical models that have a number of underlying biological parameters. Even a relatively simple model like system (2.1) has six parameters. Such a parameter space is already unimaginably large. Thus, exploring the dynamics within a given parameter range is worrisome in that we are really exploring just a tiny portion of a parameter space.

The qualitative theory of dynamical systems has been so successful because it takes advantage of the wonderful relationship between the type of model dynamics and the geometry of the underlying mathematical functions. The geometric form of the equation is, in a sense, more powerful in helping us understand the qualitative type of solution than the parameters. Discrete stock recruitment or population models, for example, produce complex dynamics for a large range of parameter values in specific mathematical models (discussed more completely in chapter 4). Despite this, all manifestations of complex dynamics in these models occur when the model has some form of over-compensatory density-dependent response (i.e., a hump-shaped function in the stock recruitment space). Thus, fluctuations are more generally understood when we inspect the geometry of the underlying model as well as its parameters. This is a good thing because it makes generalization beyond a single parameter set or range of parameters much easier.

To make this clear let us consider some familiar consumer-resource models that are common to most population or community ecology courses. The dynamics of these models are a product of their underlying geometry (which manifests, in a sense, in the isocline shapes) and less of their type and specific parameters. To see this let us examine some general well-known outcomes of a C-R model with linear consumption rates (e.g., the Lotka-Volterra-type model) and the aforementioned Rosenzweig-MacArthur model [equation (2.1)]. We will see that various manifestations of the L-V model are locally equivalent to the R-M model under different isocline arrangements. Further, both yield similar dynamic outcomes.

Figure 2.8a and a′ depicts the R-M isoclines and the L-V isoclines. The L-V model in this case has logistic growth, and so the resource isocline intersects the R axis [see Hastings et al. (1997) for details]. I have included with the R-M case a line tangent to the resource isocline (i.e., the linearization of the resource at the equilibrium). In figure 2.8a and a′ both models have similar underlying geometries; the consumer isoclines are vertical and intersect the resource isocline where it is decreasing as a function of the resource density. In such a situation these models always beget stable dynamics. While this is true, the geometry of the slope of this intersection and the proximity to the axes

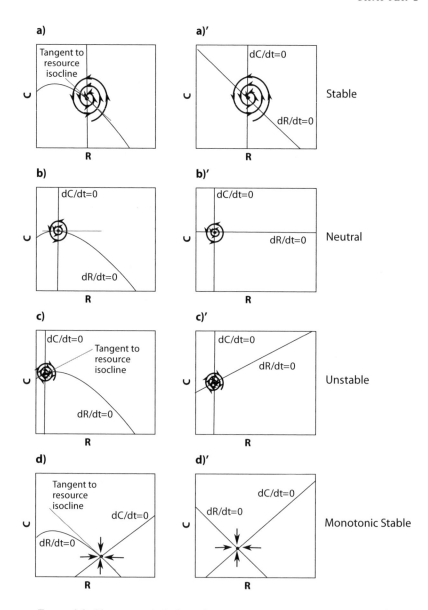

FIGURE 2.8. Phase space depictions of some common consumer-resource models. The left-hand column shows nonlinear models like the Rosenzweig-MacArthur model. Included in each figure is the tangent approximation of the nonlinear resource curve for comparison with the linear Lotka-Volterra isocline in the right-hand column. (a) and (a′) show that similar geometry from the two different models (i.e, the intersection of the isoclines is geometrically similar) produces

determine the strength of the local stability. There are clear patterns in how the strength of the local stability changes as well; we will see this in chapter 5.

Similarly, figure 2.8b shows the case where the R-M, model intersects the resource isocline exactly at the apex of its hump shape (this is where the tangent to the resource isocline is exactly horizontal). It is well known that the classic L-V model (figure 2.8b′) without logistic growth also produces an exactly horizontal resource isocline. In both cases, the local model dynamics near the interior equilibrium are neutrally stable—any parameter set that puts the isoclines in this geometry will produce such a neutral dynamic.

Let us next consider the case of the consumer isocline intersecting the resource isocline when it is increasing. In this case, the interior equilibrium is always locally unstable. This is a biologically odd case for the L-V-type formulation because growth rates are ever-increasing with increasing R density (the opposite of self-limitation). The L-V model does have a slightly different outcome here, in the sense that the solution diverges in an oscillatory fashion to infinity, while the R-M model develops a stable oscillating solution or limit cycle. Nonetheless, their similar underlying geometry points out that they both have locally unstable interior equilibrium steady states.

Considering these altogether then, we may say that any change in biology that moves us from the geometry in figure 2.8a to b and then to c tends to be destabilizing (we go from stable to neutral to unstable). This is actually what increasing a_{max} tends to do in the example in figure 2.7a. It always changes the shape of the isoclines such that we move through this geometric sequence. One can imagine creating slightly different models with similar qualitative geometric structures that all produce the same destabilization [see Rosenzweig (1971) and Fussmann and Blasius (2005) for a range of different models that do exactly this]. It should be noted that even subtle geometric differences in the models (produced either by parametric differences or slightly different geometric assumptions) may change quantitative predictions made by the different models. For example, the onset of cyclic dynamics may appear more sensitive

---

FIGURE 2.8. (*continued*) locally stable solutions. (b) and (b′) show that similar geometry from the two different models (i.e., the intersection of the isoclines is geometrically similar) produces locally neutral solutions. Another way of stating this is that the R-M model has a bifurcation when it is similar geometrically to the classic Lotka-Volterra model. Geometrically this is expected and predictable in this sense. (c) and (c′) show the case for local instability. (d) and (d′) show that local stability can be heightened when the consumer curve slants into the resource isocline (e.g., consumer interference might do this). Thus, given a geometry depicted in phase space, one can immediately intuit the local dynamics because there is a clear mapping between geometry and stability in dynamical systems.

in one model form than in another (Fussmann and Blasius, 2005). This should be no real surprise: it simply points out that different parameters and different functions can move the models through the underlying geometry at different rates.

As a final example, let us consider a deviation from the above geometry. Let us assume that the consumer isocline tilts right (e.g., the consumer interferes with itself, figure 2.8d and d'). Here, such a geometry also always remains stable (as either a nonlinear or a linear manifestation) and often tends to push the model toward even more stable dynamics (here shown as a monotonic approach to equilibrium). Although, if the consumer isocline pushes so far over that it is near the resource axis, it can become weakly stable. By virtue of being near the resource axis, the consumer is close to a bifurcation point where C goes from positive to zero values. Local bifurcations of equilibria always have zero real parts of the eigenvalue.

This weakened stability when the slope of the consumer isocline is very close to the resource axis merely reflects the fact that the consumer is so busy interfering with itself that it can barely survive. We could continue with many such examples, but hopefully the point is clear—the geometry of the mathematical functions has precise implications for the dynamical outcome. Knowing this relationship allows us to move away from a theory that relies on specific parameters to a theory that makes general statements. What this theory is doing, then, is interpreting the underlying biological assumptions in terms of the geometry. This geometry then determines the dynamics.

Importantly, we can also use such a theory to ask how modifying biological structure qualitatively changes the dynamics. An answer to this question gives us a general understanding of a problem, as it suggests that, regardless of specific parameters and perfect quantitative approaches, any biology that produces that geometry tends to yield a certain qualitative dynamical outcome. Significantly different structures frequently give different dynamical predictions: thus, this general theory, while qualitative, remains testable.

2.3 CASE STUDY: HOPF BIFURCATION IN AN AQUATIC MICROCOSM

Fussmann et al. (2000) did a detailed experimental study of an aquatic consumer-resource microcosm across a range of nutrient enrichment. They modified nutrient levels in a chemostat by varing the dilution rate, δ, and keeping track of the time series of a rotifer (*Brachionus calyciflorus*) and an algae (*Chlorella vulgaris*) under different dilution rates (figure 2.9). The *dilution rate*

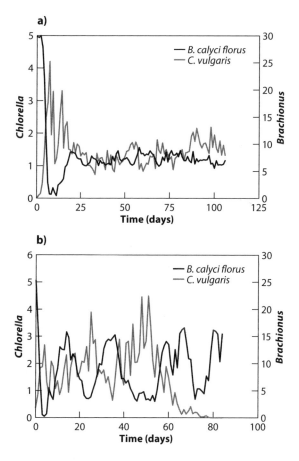

FIGURE 2.9. Example of experimental equilibrium dynamics and population oscillations for different values of the chemostat dilution rate, δ, and nitrogen inflow concentration N_i. Population dynamics of *Brachionus calyciflorus* (predator, solid symbols) and *Chlorella vulgaris* (prey, open symbols) are shown. Straight lines connect daily sampling results. Examples of trials at low $N_i = 80$ moles/liter. (a) Equilibrium at low $\delta = 0.04$ per day. (b) Cycles at intermediate $\delta = 0.64$ per day. Note different scale for *C. vulgaris* in (a). Figure adapted from data in Fussmann et al. (2000).

is defined as the rate of flow of medium over the volume of culture. Dilution rates tend to increase the production or enrichment, however, high enough values start to wash organisms out of the chemostat. Thus, at high dilution rates mortality suddenly increases. We will return to this latter aspect of the experiment in more detail in chapter 5, but here let us concentrate on the enrichment phase of the experiment.

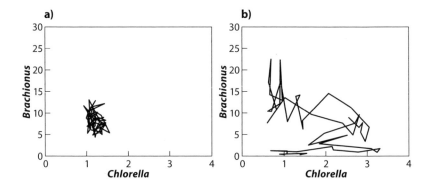

FIGURE 2.10. Example of experimental equilibrium dynamics and population oscillations plotted in the phase space. (a) Equilibrium at low $\delta = 0.04$ per day. (b) Cycles at intermediate $\delta = 0.64$ per day. Figure adapted from data in Fussmann et al. (2000).

Fussmann et al. (2000) created a basic consumer-resource model that predicted the following sequence of qualitative dynamics: (1) at low dilution rates and low nutrient levels, rotifers could not persist; (2) at moderate dilution rates, rotifers and algae reached a stable equilibrium steady state; (3) at intermediate dilution rates, rotifers and algae oscillated. Finally, with really high dilution rates, increased mortality outweighed nutrient enrichment and the dynamics crossed back over the Hopf boundary into an equilibrium steady state. Again, in chapter 5, I will discuss this latter result.

Figure 2.9a and b shows that their experiments, like the predictions from the consumer-resource model, transitioned from stable equilibrium (figure 2.9a) to oscillations (figure 2.9b) with increased enrichment. The authors, in a real sense, showed an experimental example of a Hopf bifurcation.

These results could also be shown in phase space with time implicit in the figure. Figure 2.10a depicts the equilibrium time series above (i.e., figure 2.9a) in phase space, while figure 2.10b depicts the oscillating time series [i.e., figure 2.9b, Fussmann et al. (2000)].

Fussmann and colleagues also created an experimental version of a bifurcation plot (figure 2.11) by plotting the dilution rate, δ, versus the response variable, coefficient of variation (variance/mean). The dynamics in this cumulative picture for both predator (figure 2.11a) and prey (figure 2.11b) match the story told by the individual time series. Increased dilution tends to increase variation up to the point where the dilution rate drastically increases consumer mortality (dilution rate >1 in figure 2.11). This is a simple but effective experiment highlighting that changing an important biological parameter has the

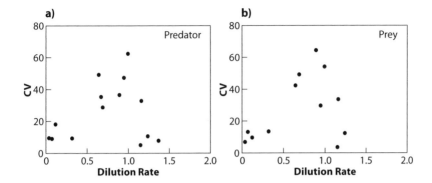

FIGURE 2.11. Relative stability of experimental predator-prey chemostat cultures. Coefficient of variation of time series (day 20 to end of trial) for all chemostat trials. Low CV indicates populations at equilibrium (CV > 0 because of noise), high CV indicates fluctuating populations. (a) *Brachionus calyciflorus* (predator). (b) *Chlorella vulgaris* (prey). Figure modified from Fussmann et al. (2000).

capability of pushing an ecological system predictably through different types of dynamical regimes.

2.4 SUMMARY OF KEY POINTS

1. There are two types of steady states (equilibrium and nonequilibrium), each of which has three qualitative different types of local dynamics (attracting, neutral, and unstable).

2. In continuous models, the real part of the maximum eigenvalue, λ, determines if the local solution is attracted to the equilibrium (negative real part), neutral (zero real part), or unstable (positive real part). Similar calculations can be used for nonequilibrium steady states like oscillating solutions.

3. The complex part of the eigenvalue determines if the solutions fluctuate near the equilibrium (if the complex part is nonzero) or simply approach monotonically (if the complex part is zero).

4. We can piece together the global dynamical picture from understanding the local dynamical behavior of all steady states (there are often many steady states in a dynamical problem).

5. Bifurcation plots are simply ways to extract information about the dynamics of a system across a range of a biological parameter. In essence, they display the results of a theoretical experiment.

6. The geometry that underlies models (e.g., manifested in the isocline shapes) give us insight into how the dynamics of a system qualitatively change (e.g., become more or less stable) when we change a biological parameter of interest. This interpretation of the geometry in essence allows one to more rigorously make statements about a biological phenomenon (e.g., an increased attack rate tends to destabilize consumer-resource interactions).

7. Simple, highly controlled microcosm experiments have shown the existence of equilibrium and nonequilibrium steady states as well as clear transitions from a "noisy" equilibrium to cyclic dynamics.

CHAPTER THREE

Of Modules, Motifs, and Whole Webs

Thus far I have motivated the problem that this book will address and laid out a mathematical primer that will aid in understanding of the chapters that follow. In the next section of the book (part 2), I proceed by considering the dynamics of important ecological modules or motifs (e.g., populations, consumer-resource interactions, food chains, omnivory). The focus in part 2, then, is to conceptually distill what aspects of population and food web structure increase oscillatory dynamics and what aspects of population and food web structure mute oscillatory dynamics. Given this, then, in the final section (part 3) I proceed by using the logic attained from this modular or motif-based theory in order to interpret the dynamics of whole food webs.

One can see that I am using the terms "module" and "motif," interchangeably. Some definitions are in order. A standard dictionary will tell you that a *module* is defined as a common interchangeable component of a system or construction designed for easy assembly. In a sense, then, a food web module is a common unit of the food web that can be used in combination with other common units of the food web (i.e., modules) with the aim of eventually assembling the whole web. Ecology is no stranger to the modular approach, as most early theory has concentrated on one of the fundamental ecological units (single species, competition, predation, mutualism). Much of this early theory thus originated from one- or two-species modules. Most people have not included single-species population models as food web modules. Here, I will consider them the base modules from which all other multispecies models depart.

Robert Paine appears to have first used the term "module" to describe a subsystem of interacting species (Paine, 1980). He considered a module a consumer and thought that its resources "behave as a functional unit" (Paine, 1980). The fairly large literature on indirect effects in food webs, those effects that propagate across several interacting species, often concentrated on smaller subsystems of tightly interacting units, although they seldom called these modules (Menge, 1995). The term seems to have been reintroduced to ecology in slightly different form by Holt (1997) in order to facilitate theoretical development beyond the well-developed theory of single species and pairwise interactions. He defined modules specifically as communities of intermediate complexity beyond pairwise interactions but well below the diversity found in most natural systems (e.g., the combination of three to six interacting species). In a sense, the modular approach Holt (1997) argued for sought to ask if we can extend pairwise ideas in a coherent fashion. Along these lines, network theorists—looking for properties in all kinds of nature's complex networks—have simultaneously considered the idea of underlying modular subnetworks (Milo et al., 2002; Camacho et al., 2007). Their terminology for this subnetwork structure is *motif*. Generally, these motifs are defined for two node/species interactions (e.g., consumer-resource interactions), three node/species interactions (e.g., food chains, exploitative competition, apparent competition), and beyond. Each size class of motif is characterized by the set of all possible interactions within that group. In this case, the single node is of little interest in quantifying network structure because it is everywhere by definition. Nonetheless, as ecologists we recognize that single nodes (populations) are important dynamically because populations can impart dynamics to the whole system (e.g., cohort cycles).

For the purposes of this book, then, I will use the term "motif" to represent all possible subsystem connections, including the trivial one-node/species case to the *n*-node/species cases. Of note, and in an attempt to delineate the different terminologies, Robert Holt personally communicated with me that he ultimately views modules "as motifs with muscles." I think this is most reasonable, as Holt's modular theory has always sought to understand the implications of the strength of the interactions on the dynamics and persistence of these units (Holt and Lawton, 1994; Holt, 1997; Holt and Polis, 1997). I will use the term "module" here, therefore, to mean all motifs that include interaction strength.

Network and graph theorists have added a great deal to our ability to categorize the structure of real food webs (Sugihara et al., 1989; Cohen et al., 1990; Milo et al., 2002). Network theory, for example, allows us to rigorously consider which motifs are common or, in the language of network theory, which motifs are over- or underrepresented in nature? Here, as in much of ecology, one compares nature to some underlying null model, and so we must remain a little cautious about what overrepresentation actually means (Artzy-Randrup and Stone, 2004; Stouffer and Amaral, 2007). Nonetheless, this technique permits us to quantify the relative presence of motifs, even allowing comparisons between different types of complex systems (Milo et al., 2002). Further, it lets use these tools too see how couplings of lower-level motifs tend to occur in natural systems (Kondoh, 2008). This is an exciting area that may allow us some traction in studying complex food web networks. It also enhances the testability of our underlying models for food web structure (Williams, 2000; Cattin, 2004; Stouffer, 2005).

Because this book is largely focused on the role of energy flux through food webs, I will hereafter use the term "module" unless otherwise indicated. In what follows, I will first move through the dynamics behind the trivial module (single populations), then move to pairwise interactions (with emphasis on consumer-resource interactions) before considering higher-order modules, and finally, consider the results from whole community interactions largely matrix-driven results that started with the contributions of Robert May, Stuart Pimm, and others (May, 1974b; Pimm et al., 1991). The role of interaction strength, or energy flux through populations, will be followed in each case with the notion that the general results on the relationship between flux through population and dynamics can be used to scale up to higher-order sets of interactions. I seek, in a simplified sense, to determine the dynamical implications of adding different modules of known interaction strengths together. The question attempts to determine whether one can understand ecological systems by understanding the dynamics of their subsystems or, rather, is the whole simply greater than the sum of its parts.

I will argue in this book that I think ecologists can understand whole systems by understanding the underlying subsystems, but this ability comes by making a concerted effort to understand how coupling different modules ultimately modifies flux within each individual module. These are the constraints that modules within a food web network place on each other. Thus, the higher-order couplings of different modules can be understood by determining if these additions excite or mute energy flows in each submodule. Others have already shown some good examples of this "addition" of modules, and I will discuss them later in the book (Bascompte and Melian, 2005; Kondoh, 2008). In part 3, I will use the theory in part 2 to interpret the results from theoretical whole-system studies. The argument will be put forward that this allows us to make sense of seemingly disparate answers and so develop a coherent food web theory. Throughout the book I will bring in data to speak to the theory; however, this is largely preliminary in scope, done to facilitate further research. I will also bring together empirical data in the last part of the book to argue for space as a critical determinant of food web structure and dynamics. I will end by discussing how human impact influences the structure and dynamics of food webs.

Part 2

FOOD WEB MODULES: FROM POPULATIONS TO SMALL FOOD WEBS

CHAPTER FOUR

Excitable and Nonexcitable Population Dynamics

While there has been a tremendous amount of theory developed for single-species population dynamics, in this chapter I choose to explore the dynamics behind basic population models with a singular lens. Specifically, I analyze population level models to highlight the general biological conditions under which population dynamics are stabilized, or destabilized, by increased population growth rates. These general results will be taken from three classes of population models. I will start with the well-known continuous logistic model before introducing the notion of discrete population models (i.e., dynamics that respond with an inherent time lag). Finally, I consider continuous models with stage-structured lags (e.g., lags due to the time taken for an individual to progress from birth to maturity). I will show that increasing per capita growth rates tend to stabilize population models when the underlying dynamics are monotonic (i.e., show no evidence of fluctuations). However, lags in population models tend to introduce dynamics with oscillatory decays to equilibrium or sustained oscillations around the carrying capacity, K. Such oscillatory decays or sustained oscillations are only further destabilized by increased growth or production rates. This simple result, from a general class of population models, will lay the foundation for interpreting the dynamics and stability of food webs in later chapters.

4.1 CONTINUOUS RESOURCE DYNAMICS

Let us start with the continuous logistic growth model:

$$\frac{dR}{dt} = rR\left(1 - \frac{R}{K}\right), \tag{4.1}$$

where r is the intrinsic per capita rate of increase and K is the carrying capacity.

The logistic is a simple model, but one that qualitatively captures a lot of real population dynamics, at least for some period in time. Laboratory microcosms—population dynamics in a vacuum so to speak—often are well approximated by the logistic (Gause, 1932, 1934). Numerous invasions follow population trajectories akin to the logistic (Fritts and Rodda, 1998; Forsyth and Caley, 2006). Invaders often arrive to find high resource densities and bewildered predators, and so the early biology follows some of the necessary requirements for logistic growth. It is this empirical success that makes such a model an excellent foundation for more complex interaction models. Although an incredibly simple equation, it phenomenologically embodies the dynamics of populations in isolation.

At low population densities, the resource population, R, effectively grows exponentially [equation (4.1) is well approximated by $dR/dt = rR$ because R/K is near zero] but as the population increases toward the carrying capacity, K, the population growth rate approaches 0 [i.e., the dynamical system equation (4.1) has an equilibrium at $R^* = K$]. Using the tools discussed in chapter 2, one can determine the local stability by calculating the eigenvalue, λ, at the equilibrium $R^* = K$. After linearizing and substituting $R^* = K$ into equation (4.1), λ is found to be determined by the following relationship:

$$\lambda = -r. \tag{4.2}$$

Mathematically this means that the local model approximation can be written as

$$\frac{d\epsilon}{dt} = \lambda\epsilon = -r\epsilon, \tag{4.3}$$

which has the solution

$$\epsilon(t) = \epsilon(0)e^{\lambda t} = \epsilon(0)e^{-rt}. \tag{4.4}$$

Because $-r$ is always negative, the population dynamics near the equilibrium ($R^* = K$) are always locally stable, and the strength of this stability is determined by $-r$. The larger r is in absolute value, the more rapid a perturbation decays back to the equilibrium because a larger negative r causes the exponential term to recoil back to zero more rapidly. The continuous system is thus stabilized by increasing per capita growth rates, r, but is uninfluenced by the carrying capacity, K. Note that this continuous model produces only monotonic dynamics (figure 4.1). That is, it has no ability to produce oscillatory decay or oscillation in the dynamics.

Here, we come across a theme that will weave through this book: Population dynamic models that produce monotonic dynamic trajectories tend to increase in stability with increased production. Recall that an equilibrium is a dynamical entity in that there is a continuous turnover of individuals throughout a population. The population simply remains at carrying capacity because the

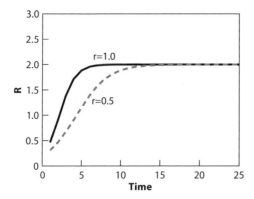

FIGURE 4.1. With continuous logistic growth, increasing r simply speeds up the rate of return of a pertubation to the equilibrium (i.e., increases the resilience). The dynamics are always monotonic trajectories and so never display overshoot dynamics. $r = 0.50$ for the dashed line, and $r = 1.0$ for the solid line ($K = 2$).

flux of new individuals into the population equals the flux of old individuals out of the population. As such, r is directly responsible for determining the flux through the population, R, near the carrying capacity, K. Even if K grows, the flux in and out of the population at $R^* = K$ remains at r. The more rapid the flux in, the more rapid the population can respond to a local perturbation.

Figure 4.1 shows two trajectories to the same carrying capacity from the same initial resource density (i.e., the trajectories start at the same R value at time $= 0$). Not surprisingly, the population with the higher r (solid curve) rises up and approaches carrying capacity more rapidly then the low-r population. In a sense, then, a perturbation that knocks both population densities down to the same initial value would find that the high r population returns to the equilibrium quicker. If the world remained this simple, such that all populations had monotonic trajectories, our understanding of stabilizing mechanisms would be fairly trivial. Anything that increased a population's intrinsic growth rate would similarly drive heightened stability. Unfortunately, it seems that oscillating dynamics appear to be relatively common in one form or another (an empirical issue discussed later), and so this simple model, although excellent in a completely isolated sense, frequently fails in a more complex biological scenario. I now turn to some results from population dynamics that argue that oscillations, or nonmonotonic dynamics, readily appear in population dynamics.

In what follows, I will refer to dynamics that are monotonic as nonexcitable, and dynamics that have oscillations of any sort (either permanently or on the

Box 4.1
EXCITABLE AND NONEXCITABLE DYNAMICS

Definition 4.1 Nonexcitable Population Dynamics: dynamic trajectories that monotonically approach an equilibrium after a localized perturbation. In a simplified sense, these are trajectories that show absolutely no sign of fluctuation (see figures 4.1 and 4.2a for examples).

Definition 4.2 Excitable Population Dynamics: dynamic trajectories that approach an equilibrium in an oscillatory manner (figure 4.2b) or approach an oscillating steady state (figure 4.2c).

■

approach to the equilibrium) as excitable (Box 4.1). I invoke this terminology because I will argue in this chapter, and throughout this book, that these underlying dynamical characteristics have implications for the way changing biological parameters act on the local stability of steady states. Specifically, I will show that oscillatory decay or sustained oscillations are easily excited to display even greater fluctuations by increased rates of energy flux through a population, whereas nonexcitable, or monotonic, population dynamics are stabilized by increased rates of flux through a population, as seen above for the continuous logistic. Increased growth rates in this nonexcitable model merely ramp up the population's ability to return to the steady state after a perturbation [equation (4.1)].

4.2   FROM NONEXCITABLE TO EXCITABLE POPULATION DYNAMICS

Until this point we have assumed that population dynamics respond instantaneously to changes in their own density (e.g., dR/dt follows instantaneous changes in rates). This is unlikely to always be the case. The density of organisms at a specific point in time can have strong dynamical consequences for the population at some distant time in the future (Durrett and Levin, 1994). As an example, the density of adults in the breeding season may have important ramifications for the number of offspring recruited into the population the next year. Fisheries ecologists have recognized the dynamical implications of this lagged recruitment since the seminal contributions of W. E. Ricker (1954). It is therefore necessary to consider how this biological structure (i.e., a lagged population response) influences population dynamics.

I now turn to a class of models called discrete equations. These models have an implict lag in them. For consistency, I am going to consider a discrete single-population equation that has many of the biological properties of the logistic equation. That is, I will employ the following discrete population equation known as the Ricker model (Ricker, 1954):

$$R_{t+1} = R_t e^{\beta(1-R_t/K)}, \tag{4.5}$$

where β is the discrete analog to the continuous rate of the increase parameter, r, and K is the carrying capacity. Again, the model has an implicit lag of 1 time unit between successive model predictions. The time unit here is quite open and can represent any relevant time scale, however, it is often 1 year, representing our tendency to monitor populations on a yearly basis. Given a yearly time interval, equation (4.5) describes next year's population density given this year's population density. This lag, as lags often do, makes for much greater dynamical instability [e.g., May (1976); Hastings and Wollkind (1982); Hastings (1983)].

To see this, I now briefly explore the local stability properties of equation (4.5). As in continuous models, the local stability of this model at the equilibrium $(R^* = K)$ can be determined by differentiating the equation at the equilibrium (i.e., determining the linear term at the equilibrium) with respect to R,

$$\frac{dF_R}{dR} = \left(1 - \beta\frac{R}{K}\right) * e^{\beta(1-R/K)}, \tag{4.6}$$

and substituting in the equilibrium $(N^* = K)$. The eigenvalue, λ, therefore can be determined to be

$$\lambda = (1 - \beta). \tag{4.7}$$

Again, as in continuous models, λ tells us how the dynamics near the equilibrium behave. Mathematically, the linearization for this discrete model can be written as

$$\epsilon_{t+1} = \lambda\epsilon_t, \tag{4.8}$$

where ϵ describes the dynamics of the perturbation extremely close to the equilibrium $(R - R^* = \epsilon)$. The form of the linear dynamical model here is different from that of the continuous model because it clearly remains a discrete representation. Because of this, the implications for local stability of the eigenvalue, λ, are slightly different.

Notice that if $|\lambda|$ is greater than 1, the perturbation in equation (4.8) grows with time (ϵ grows with each recursion). This is independent of the sign of the real part of λ because a large negative or positive value makes the perturbation grow, albeit with qualitatively different dynamics that I will discuss shortly. Thus, $|\lambda| > 1$ implies an unstable equilibrium. On the other hand, if $|\lambda < 1|$, the perturbation decays and the equilibrium is stable, while $|\lambda = 1|$ is the neutral case. The case of $\lambda = 0$ is the case of maximal stability, or superstability, because the perturbation is perfectly eliminated in one time step. Hence, equation (4.7) determines the local stability of the model at $R^* = K$, and the local stability depends solely on the per capita growth rate, β (recall that the continuous model depended solely on the continuous growth rate, r). As β is increased from 0 to 1 the system goes from neutrally stable (i.e., $\lambda = 1$) to maximally stable ($\lambda = 0$). Thus, the local stability increases as β increases from 0 to 1). This part of the result resonates with the continuous model.

However, as β increases beyond 1, the system becomes less stable (i.e., the distance of the eigenvalue, λ, from a maximally stable zero value suddenly increases at this point). Curiously, this change in effect (i.e., from a stabilizing to a destabilizing effect of growth rate) occurs precisely at the point the eigenvalue, λ, switches from positive to negative. This negative eigenvalue implies that the dynamics suddenly overshoot the carrying capacity and so have a tendency toward sustained oscillation or, at the least, oscillatory decay to the equilibrium. Either way, in our new terminology the dynamics of the population have become excited. To see this, notice that equation (4.8) describes the local dynamics near the equilibrium. Because of this transformation near the equilibrium, the solutions can be negative (i.e., $R > R^* = K$) or positive (e.g., $R < R = K^*$). If the pertubation started off negative, within 1 time unit it is on the positive side of the equilibrium [a negative $\epsilon(t)$ times a negative λ makes the next solution in equation (4.8) positive]. One can carry on such iterations through time to see that the solution bounces back and forth across the equilibrium, approaching the equilibrium if $|\lambda|$ is between 0 and 1, or departing away from the equilibrium in a fluctuating fashion if $|\lambda|$ is greater than 1.

Figure 4.2a shows a monotonic approach to the carrying capacity that occurs when the eigenvalue is less than 1 and positive. On the other hand, figure 4.2b and c show two time series for a stable negative eigenvalue ($|\lambda| < 1$) and an unstable negative eigenvalue ($|\lambda| > 1$), respectively. In both cases, the dynamics have started to flip back and forth around the equilibrium. Thus, from a biological point of view, the negative eigenvalue means that the dynamics overshoots the equilibrium and so no longer monotonically approaches the equilibrium. Thus, increasing the population growth rate by increasing β creates the unfortunate problem of eventually driving excitable dynamics. Hence, what was once stabilizing (faster intrinsic growth rates)

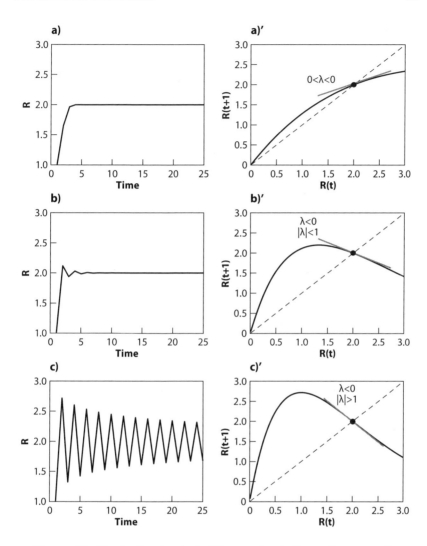

FIGURE 4.2. Simple discrete map that is an analog to the continuous model (equation 4.1). The model shows both monotonic and excited (overshoot) dynamics. (a) Low β values (e.g., $\beta < 1$) produce monotonic trajectories akin to the continuous model above. (b) Moderate β values ($\beta = 1.5$) produce nonmonotonic, or excitable, trajectories with oscillatory decay to equilibrium. (c) Still higher β values ($\beta = 2.0$) produce nonmonotonic excitable trajectories that permanently oscillate. (a')–(c') give the corresponding state space representation of equation (4.5) for the specific parameters $\beta = 0.50$, $\beta = 1.50$, and $\beta = 2.0$, respectively. $K = 2$.

now becomes a destabilizing biological factor when population dynamics can overshoot.

I have drawn beside each of the time series in figure 4.2a–c the accompanying state space representation, which allows us to look at the shape of equation (4.5) given the parameter values (solid curve in figure 4.2a'–c'). I have also added the one-to-one line (dashed line) on this state space. For discrete single-species models the intersection of these two lines gives us the equilibrium (solid dot in figure 4.2a'–c'). In this state space graph, the slope at the equilibrium is λ [see Yodzis (1989) for an excellent tutorial on these graphs]. From this, one can immediately see that λ is small and positive in figure 4.2a' (meaning stable and monotonic), relatively small and negative in figure 4.2b' (meaning stable with oscillatory decay), and large and negative in figure 4.2c', thus driving unstable oscillatory dynamics. Again, the underlying geometry of the model is responsible for the dynamics of these mathematical models. A variety of different functions and parameter sets with similar shapes will all produce the same sequence of dynamical events. Note as well that increasing the per capita growth rate, β, increases the nonlinearity of the problem. In essence, it makes the hump shape more pronounced, thus inspiring excitable dynamics.

The dynamics of a single-species model without some sort of delay are always monotonic or nonexcitable. Lagged population models, on the other hand, tend to readily show excitable dynamics relative to continuous models without lags. I will show that this result, in many different guises, arises frequently in ecological interactions. The lag in this population model can stand for an enormous amount of biology. Lags can be due to age structure in a population, a response to prey, a response to predators, and so on. One might expect that this result therefore ought to readily manifest in more explicitly modeled scenarios that produce lagged responses.

A simple thought experiment illustrates how a lag and a high r drive fluctuations. Consider a situation where you are in a room that is cold and the thermostat on the wall has no numbers, just an up-and-down knob. If you stay in the room and continuously adjust the the thermostat, depending on whether you feel too warm or too cold, you will relatively quickly aproach a comfortable temperature. This is the continuous model, and one expects a relatively well-behaved approach to a comfortable temperature. On the other hand, if you change the temperature and then leave for a whole day (i.e., incur a time lag), you may return to an extremely warm room the following day. If your tendency in such a case is to adjust with bravado, you will find that you will reduce the temperature and the next day it will be quite cold. If you continue with this experiment, changing the thermostat wildly on a daily basis, the room will fluctuate between extreme cold and extreme warmth. If you instead adjust the thermostat gently, you may not produce any overshoots and likely will find a comfortable temperature after some time. The adjustment level in this thought experiment is akin to β above, while the day in between temperature adjustments is the lag.

The above results point out that increasing growth rates (i.e., β) ought to eventually drive instability in the face of a lag, and this turns out to be generally true. Robert May showed this with a number of simple discrete models, while emphasizing that simple models with discrete lags readily generate complex dynamics (May, 1976). The above exercise has shown us that a population behaves differently, in terms of whether it is stabilized by increased growth rates or not, when the dynamics are monotonic or excitable. In what follows, I will argue that this simple distinction allows us to organize a group of seemingly disparate stability results from the literature, and taken within this framework offers a coherent theory about the stability of ecological systems.

Population dynamics theory has moved beyond logistic population growth to consider the influence of age and stage structure (i.e., life stages are modeled separately). Because the lags implicit in discrete models make the dynamics less stable on average, one can expect that continuous models with age- or stage-structured lags may do the same. I now consider whether the above results on stability for the continuous logistic are changed by the inclusion of stage structure, a ubiquitous aspect of all populations.

4.3   STAGE-STRUCTURED RESOURCE DYNAMICS

From our previous work on discrete models, it was found that increasing production (e.g., β) first stabilizes and then destabilizes population dynamics. To interpret this result I conjectured that the stabilization occurred only when the dynamics were monotonic. Given monotonic, or nonexcitable, population dynamics, increased flux through a population simply acts to return it to a given equilibrium density more rapidly. On the other hand, if the dynamics are oscillatory in any form, such an increase in production has the consequence of further exciting overshoot dynamics and delaying the actual return time to the equilibrium. One may expect the same qualitative result from a stage- or age-structured model. To approach this, I extend the continuous logistic to the following delayed logistic model:

$$\frac{dR}{dt} = rR \left(1 - \frac{R(t - \tau)}{K} \right), \tag{4.9}$$

where r is the intrinsic rate of increase, K is the carrying capacity, and τ is the delay incurred by some aspect of stage structure (e.g., recruitment from adult biomass τ years ago).

The analysis of delayed models is similar in spirit to the analysis of continuous models discussed in chapter 2, with the addition that the characteristic equation that determines the eigenvalue is a transcendental equation. Transcendental equations are often not analytically tractable, and as such the eigenvalues must be determined via numerical techniques. Despite the trickiness of transcendental equations, there are some excellent mathematical contributions presenting delayed differential equations that employ qualitative analytical techniques to delineate regions of stability (Bellman and Cooke, 1963; May, 1974a).

The eigenvalue, λ, for the delayed logistic is determined from the characteristic equation for $P(\lambda)$:

$$P(\lambda) = \lambda + re^{-\lambda\tau} = 0. \tag{4.10}$$

In order to see how a stage-structured resource model responds to increased intrinsic growth rates, I proceed by varying r and solving numerically for the eigenvalue of equation (4.10). As before, I find that the resource dynamics are stabilized in the monotonic region of parameter space (i.e., the eigenvalue is real, and so the dynamics are monotonic near the equilibrium) and destabilized when the dynamics enter the excitable region of parameter space (figure 4.3; eigenvalues are complex).

This result is consistent with the more qualitative analytical techniques employed by May (1974a) that find the sequence of dynamics in a number of delayed resource models (e.g., the delayed Gompertz model) goes from montonically stable to oscillatory decay to sustained oscillations, with increased r. In May's paper, though, the relative stability is not actually determined (i.e., whether r stabilizes or not within the different regions). Rather, these early results tended to focus on the fact that the actual destabilization, or transition to dynamic oscillations, is a function of both r and the delay, τ. The general result is that the greater the delay, the less the intrinsic growth rate, r, required to drive a transition into the oscillatory domain (May, 1974a). It is important to the thesis of this book that we emphasize that, regardless of the magnitude of τ, increasing the intrinsic growth rate from 0 causes r to always drive a sequence of stabilization followed by destabilization—the duality of how growth rates influence stability remains true. Further, this switch in the influence of r on stability is intimately associated with the qualitative type of dynamics (monotonic versus excitable) as stated above.

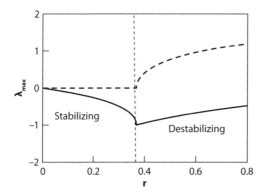

FIGURE 4.3. Eigenvalue response of a continuous delayed logistic model as r is increased. The real part of the eigenvalue, which determines stability, is the solid curve, and the complex part of the eigenvalue, which determines whether the solution shows overshoot dynamics or not, is the dashed curve. Notice that the stabilizing region of r corresponds to a zero imaginary part (monotonic dynamics), while the destabilizing phase again corresponds to the nonzero imaginary (excitable) region of the eigenvalue. Parameter value: $\tau = 1$.

4.4 EMPIRICAL EVIDENCE FOR EXCITABLE DYNAMICS

The theoretical results from the previous sections make it important to assess whether most real population dynamics are excitable (i.e., oscillatory decay or sustained oscillations) or monotonic. Where cycles have been analyzed, oscillations appear to commonly occur in populations with specialized predators (Turchin et al., 2003). There is also abundant evidence of oscillatory signatures in simple microcosm studies (McCauley et al., 1999; Fussmann et al., 2000). Nonetheless, while there is evidence of the existence of oscillatory signatures in real populations [e.g., Hanski et al. (1991); Turchin et al. (2003)], there has been less concern about the relative amounts of qualitatively different types of dynamics. In a 1998 paper, Bruce Kendall and colleagues sought to find out whether there were large-scale macroecological patterns in population cycles and so concerned themselves with the relative frequency of cycles in time series data from empirical investigations (Kendall et al., 1998). They found that, on average, approximately 30 percent of 700 time series from the Global Population Dynamics Database exhibited cyclic dynamics. Further, there was considerable variance around this average over latitudinal gradients for mammals such that mammals exhibited cycles in excess of 50 percent of the populations taken from northern clines (figure 4.4).

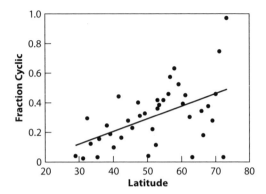

FIGURE 4.4. Percentage of mammals from a given latitudinal area that show evidence of cyclic population dynamics. Taken from Kendall et al. (1998).

Of all species, birds were the most comprehensively represented, and in birds population cycles seemed evenly distributed over all taxonomic orders. Here too, cycling seemed to occur in bird species that had a number of different trophic roles, from predatory species to insectivores and granivores (Kendall et al., 1998). The early data on the relative frequency of cycles suggest that population cycles are relatively ubiquitous. This estimate of 30 percent is likely a conservative estimate of populations that are excitable in that the analysis sought evidence of cyclicity and so may have overlooked time series that showed only modest evidence of decaying oscillations (quasiperiodicity). At the very least, it appears as if nature's complex systems harbor both qualitative types of population dynamics.

In a similar fashion, Forsyth and Caley (2006) pooled the dynamics of seven large herivores in order to determine whether large herbivores displayed evidence of population overshoot dynamics (i.e., evidence of excitable dynamics) after introduction into a new range or release from harvesting. The dynamics of six of the seven populations showed strong evidence for overshoot dynamics, suggesting that, at least among herbivores, either oscillatory decays or sustained oscillations seem potentially widespread. This large herbivore literature refers to this same overshoot of carrying capacity as "irruptive" dynamics. More recent evidence from Kaji et al. (2009) found that two different populations of sika deer showed evidence of excited dynamics, with one showing decaying oscillations and another showing sustained oscillations after introduction.

These results leave us wondering whether human-driven change in ecological systems due to harvesting, fragmentation, and diversity loss will increase the likelihood and magnitude of population fluctuations. A recent paper by

Hsieh et al. (2006) examined the variability of exploited versus unexploited fish species and found that, indeed, exploited species tended to show evidence of increased variation. Anderson et al. (2008) followed up on this study and examined several different plausible hypotheses for this increased variability. The data strongly support the notion that fisheries truncate age structure and simultaneously alter a number of critical aspects of fish life history, promoting increased population growth rates. Much fisheries literature has found that fishing pressure changes organismal life history by increasing juvenile growth rates and decreasing age at maturation (Jennings et al., 1999). Consistent with this, Anderson et al. (2008) found that this increased growth rate, as discussed in this chapter, acted to increase the nonlinearities in the dynamics and in doing so excited the population dynamics into greater fluctuations.

4.5   SUMMARY: THE DUAL NATURE OF POPULATION GROWTH RATES

1. All resource models that beget monotonic dynamic approaches to the equilibrium carrying capacity, K, are stabilized by increased growth rates, r. This makes intuitive sense because a perturbation returns at a rate proportional to the population growth rate as long as the population growth rate has no negative consequence (i.e., a negative consequence is that a large growth rate can cause the population to overshoot the carrying capacity, thus delaying return time).

2. All resource models that beget oscillatory dynamics or oscillatory decay to the equilibrium are destabilized by increased growth rates. Thus, even though a population rapidly increases back toward the equilibrium, its excessive growth rate overshoots the equilibrium and ultimately weakens the local stability.

3. Lags of any form (e.g., discrete models, stage-structured models) are consistent with (1) and (2) but are more likely dominated by the destabilizing aspects of increased intrinsic population growth rates because these population models readily produce more complicated dynamics (i.e., oscillations and complex dynamics).

4. A preliminary study by Kendall et al. (1998) of a large data set of time series suggests that oscillatory population dynamics are relatively common in nature (approximately 30 percent of 700 natural time series showed significant evidence of cycling). Similarly, a synthetic review of large herbivore

dynamics found that the potential for oscillatory decay or sustained oscillations is widespread.

5. Anderson et al. (2008) found that exploited fish were more variable than unexploited fish and that this appears to be due to the fact that fisheries pressure tends to truncate age structure, decrease age at maturation, and increase somatic growth rates that in sum lead to higher population growth rates. The increased intrinsic growth rate acts to increase the nonlinearities in stock recruitment dynamics and so excites the population dynamics to greater fluctuations.

CHAPTER FIVE

Consumer-Resource Dynamics: Building Consumptive Food Webs

The consumer-resource interaction is one of the fundamental building blocks of food webs. Consumption interactions shunt energy and nutrients through ecological networks and play a critical role in the functioning of all eco-systems. This chapter is consistent with the last chapter in that the intention is to glean general rules from theory. It is not a comprehensive review of C-R theory, rather I will consider C-R theory within the duality laid out in the previous chapter such that I will ask how C-R systems that are nonexcitable and excitable respond to changes in interaction strength. For an excellent, thorough account of C-R theory, see the book by Murdoch et al. (2003). Before I consider C-R theory, though, it is necessary to take a brief excursion into the minefield of interaction strength measures. I will then look at continuous C-R models to argue that, despite the fact that some models produce a richer array of dynamics, most give the same qualitative answers to changes in interaction strength and so contribute to a general theory. Given these general patterns, I highlight two underlying mechanisms that I conjecture are behind the stabi-lization of C-R interactions. Finally, I end by briefly exploring some empirical results that speak to this general C-R interaction theory.

5.1 INTERACTION STRENGTH

At first glance, there appears to exist a bewildering variety of interaction strength metrics (Berlow et al., 2004). Luckily, the idea of interaction strength is fairly intuitive, and so beneath all these definitions lie a few relatively general types of measures (Laska and Wootton, 1998; Wootton, 2005).

Here, for simplicity, I break the definitions into two classes. One class of interaction strength metrics attempts to explore instantaneous rates of change in one species with respect to another species. These metrics of interaction strength are closely aligned with theoretical models that follow continuous rates of change (i.e., dN/dt). These metrics define rates of change in one species with respect to those in another species.

There is also a second class of interaction strength metrics that follows the longer-term influence of the removal of (or change in) one species on the density of another focal species. These longer-scale definitions stem from the pioneering experimental treatment of interaction strengths by Paine (1992) but include similar theoretically derived defintions such as the inverse Jacobian (Yodzis, 1988). Recall that the Jacobian, defined in chapter 2, is simply the linearization of a mathematical model presented in matrix form. This second class, by virtue of its longer time scale and experimental manipulation, allows for the long-term influence of all indirect pathways [see Wootton (2005) for a good discussion of this issue]. This second class of metrics relies on the assumption of attaining equilibria and so defines interaction strength based on changes in the equilibria of a focal species. This estimate does not consider time and so is not a rate; this is a significant difference from the short-term rate-based estimates above. I will call this second class of metrics *functional interaction strength* because they have been used empirically to measure the long-term influence of one species on another.

In a synthetic contribution on interaction strength, Wootton (2005) pointed out that a relatively common metric of interaction strength measures the short-term influence of 1 unit of a given species on a unit of another species. Two metrics of interaction strength have permeated the literature, and these instantaneous measures of interaction strength (classically denoted a_{ij} for the effect of species j on speces i) are as follows:

(i) **Interaction Strength (May)**: the population's growth rate $[dN_1/dt = F_1(N_1, N_2)]$ relative to the change in another species' density (mathematically defined as $a_{12} = \partial F_1/\partial N_2$). At equilibrium this is the classic interaction strength matrix of May (1974b) and others, and

(ii) **Per Capita Interaction Strength (Levins)**: this measure depends on the population's per capita growth rate, f [i.e., $f_1 = F_1(N_1, N_2)/N_1$], relative to the change in a species (mathematically defined as $a'_{12} = \partial f_1/\partial N_2$). This latter method is attributed to Levins (1968).

The above definitions are easily applied to intraspecific cases, and so both methods ultimately supply us with a matrix of interaction strengths. At first glance, there are notable differences between these metrics, although if we transform the community matrix of Levins (1968) by multiplying the N_i

back into the matrix (e.g., let $a_{12} = a'_{12} \times N_1$), we actually return to a matrix identical to May's interaction matrix under the assumption of equilibrium.

The Jacobian interaction strength matrix employed by May (1974b) is a direct line to interpreting local stability. For this reason alone, I will tend to employ the Jacobian matrix at equilibirum as a measure of interaction strength, recognizing that all results could easily be interpreted in terms of the Levins measure of per capita interaction strength if biomass/density is known.

5.1.1 THE RELATIONSHIP BETWEEN MAY'S INTERACTION STRENGTH AND MATERIAL FLUX

One nice aspect of rate-based interaction strengths is that there is a relatively straightforward mapping between the above interaction strength measures and important biological aspects that govern the flux of material between the interaction. Further, data exist to estimate these underlying parameters in order to facilitate comparison between theory and emprirical results. The interspecific interaction strengths, for example, are a function of attack rates, a, and the biomass conversion efficiency, e. These parameters can be estimated from time series data, experimental trials, and physiological considerations (Wootton, 2005).

To illustrate May's interaction strength and its relationship to biological flux rates within and between organisms, let us start by considering the well-known Rosenzweig-MacArthur C-R model (Rosenzweig and MacArthur, 1963):

$$\begin{cases} \dfrac{dR}{dt} = rR\left(1 - \dfrac{R}{K}\right) - \dfrac{a_{max}RC}{R + R_o}, \\[3mm] \dfrac{dC}{dt} = \dfrac{ea_{max}RC}{R + R_o} - mC, \end{cases} \tag{5.1}$$

where R and C are resource and consumer, respectively, r is the per capita rate of increase, K is the carrying capacity, a_{max} is the maximum attack rate, R_o is the half-saturation density, e is the fraction of biomass consumed and turned into consumer biomass, and m is the per capita mortality rate of the consumer.

In this case, May's interaction strength matrix at equilibrium is

$$\begin{bmatrix} -rR^*/K + \dfrac{rR}{R + R_o}\left(1 - \dfrac{R}{R + R_o}\right) & -\dfrac{a_{max}R^*}{R^* + R_o}\left(1 - \dfrac{R^*}{R^* + R_o}\right) \\[4mm] \dfrac{ea_{max}C^*}{R^* + R_o}\left(1 - \dfrac{R^*}{R^* + R_o}\right) & 0 \end{bmatrix}. \tag{5.2}$$

Upon examining the interaction strengths [equation (5.2)], it becomes clear that we can attempt to relate the interspecific terms and the intraspecific terms

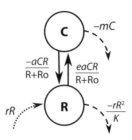

FIGURE 5.1. Box flow diagram for the R-M model. Note that key parameters alter biomass flux through this interaction (a_{max}, e, R_o) or modify the biomass growth (r) or loss terms (r, K, and m). The solid arrows denote the coupling terms, while the dashed arrows denote the loss terms and the dotted arrows the resource uptake term (rR). The flux rates allow us to interpret the dynamical response of a C-R interaction from the perspective of how biological parameters alter the fate of biomass flux in the interaction.

to key parameters that govern the material flux between a consumer and its resource. Given that there is a clear relationship between interaction strength and flux rates, any term governing the strength of this flux between consumer and resource may be used as a surrogate for interaction strength. The job of estimating the precise interaction strengths clearly requires the additional information of population density.

To facilitate this mapping between May's interaction strength matrix and flux rates, I start by breaking down the traditional consumer-resource interaction into a biomass, or material, flux diagram (figure 5.1, as depicted for the R-M model). This diagram shows that there are three types of terms that govern the fate of biomass flux through this interaction: (1) coupling terms (solid arrows that join nodes, figure 5.1); (2) dissipative loss terms (dashed arrows leaving a node but not attached to another node, figure 5.1), and (3) resource uptake terms (the rR term, dotted arrow). The dissipative loss term, hereafter referred to as the *loss term*, encompasses both linear losses (e.g., $-mC$) and self-damping losses ($-rR^2/K$).

I will argue in the next section that the dynamical response of the C-R interaction to variation in any parameter depends on the strength of the coupling terms (hereafter referred to as *coupling strength* in order to distinguish it from interaction strength) relative to the strength of the loss terms. Note that the coupling terms, or coupling strength, can be different from May's interaction strength because interaction strength often includes loss terms as well. The coupling strength definition always cleanly separates the coupling terms from the loss terms.

The major result of this chapter will be to show that increases in the flux of material through the coupling terms relative to the loss term tend to destabilize an excited C-R interaction. From an interaction strength perspective, this is akin to stating that as the interspecific interaction strengths of the C-R interaction matrix (nondiagonals) grow relative to the intraspecific interaction strengths (diagonals), already excited dynamics can be expected to become less stable.

5.2 CONSUMER-RESOURCE INTERACTIONS: TWO QUALITATIVE RESPONSES TO CHANGES IN A PARAMETER

Although I examine the implications of consumer-resource interactions largely using the Rosenzweig-MacArthur model [equation (5.1)], other models could just as easily illustrate the points laid out here. To emphasize this, below I also briefly discuss how the qualitative results apply to the Lotka-Volterra model with logistic growth. Nonetheless, the R-M model is biologically well accepted and produces a range of dynamical outcomes (e.g., stable equilibrium to limit cycle behavior) and therefore serves well as a model system. After the analysis presented here, I will consider these results within the broader set of C-R models.

Luckily, one can understand how the various biological parameters in the R-M model influence stability by considering how parameters change the *relative* position and shape of the isoclines (see chapter 2). It turns out that there are really only two qualitatively distinct ways the parameters of the R-M model change the shapes of the consumer and resource isoclines. First, the most common geometric response to changing parameters in this model is that the parameter changes the position of the consumer isocline relative to that of the resource isocline. Figure 5.2a–d shows an example of this. Notice that the consumer isocline progressively shifts to the left of the resource isocline in this example (figure 5.2a–d).

This relative shift in isocline can manifest in a variety of ways. Some parameters, for example, simply move the consumer isocline and the resource isocline is not modified (e.g., *e*). Other parameters change only the resource isocline (e.g., *K*), while still others change both isoclines. Regardless of how this happens, a number of parameters generally drive this change in the isoclines. As an example, increasing a_{max} alters both isoclines but still tends to move the consumer isocline to the left of the resource isocline. Because most parameters move the isoclines in this manner, it turns out that most parameters

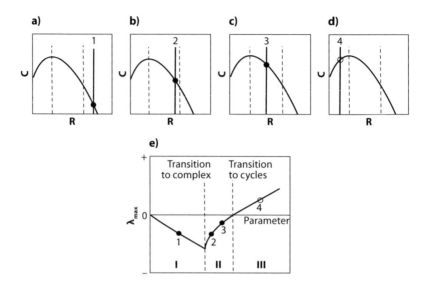

FIGURE 5.2. Qualitative response of consumer-resource isoclines to variation in all but one parameter in the R-M model. Specifically, increasing a_{max}, K, or e drives a relative change in the isoclines that matches (a)–(d) and produces the eigenvalue, λ, response shown in figure 5.2e. Similarly, decreasing m or R_o changes the isocline positions as depicted in (a)–(d) and so also produces the eigenvalue response shown in (e). Note that increases in flux through the C-R interaction, or similarly, decreases in the natural loss terms in R and C result in this dynamical sequence of events. The rightmost dashed line identifies the consumer isocline position that marks the transition to oscillatory decay (eigenvalues become complex), and the leftmost dashed line marks the isocline position that marks the transition to sustained oscillations or cycles. The three regions these transition curves separate are identified as I, II, and III in (e). Solid circles represent equilibria that are locally stable, and the open circle represents a locally unstable equilibrium.

have a similar effect on C-R dynamics. We will see, though, that some have to be increased to move from figure 5.2a to figure 5.2d (e.g., e, a_{max}, K), and others have to be decreased (R_o, m).

The only parameter that does not operate to move the relative position of the consumer isocline as in figure 5.2a–d, r, instead tends to change the slope of the intersection point of the two isoclines. Recall that the intersection is the equilibrium. Not surprisingly, this different geometric influence drives a different dynamical response to changes in this parameter. Below, we explore these two cases one at a time and discuss the general ecological significance of these results.

5.2.1   EXCITING CONSUMER-RESOURCE INTERACTIONS (a_{max}, e, K, R_o, m)
In this section I show that when a parameter is changed such that it increases
the amount of biomass flux between the consumer and resource *relative* to the
size of the loss terms, the interaction tends to be destabilized.

I want to emphasize what relativity means here. A relative increase in
the coupling strength does not necessitate that the coupling terms increase in
strength. It is also possible that the coupling strength increases relative to the
strength of the loss terms if the loss terms are reduced. It turns out that when the
problem is posed this way, almost all the parameters (i.e., a_{max}, e, K, R_o, m)
in the R-M model have a similar effect—all of them act to change the relative
strength of the coupling terms.

Figure 5.2a–d depicts the changes in the consumer isocline (solid vertical
line) relative to the humped resource isocline when one of the above biological
parameters is varied in a way such that it increases the relative flux through the
coupling terms. As one moves from figure 5.2a to figure 5.2d the consumer
isocline moves from right to left relative to the resource isocline. The two
dashed lines in these figures are important and represent:

(i) the rightmost dashed line (figure 5.2) represents the transition from
     monotonic to oscillatory decay dynamics, and
(ii) the leftmost dashed line (figure 5.2) represents the transition from
     oscillatory decay to sustained oscillatory dynamics or cycles (often
     called the Hopf bifurcation). This dashed line always occurs at the
     peak of the resource isocline in the R-M model.

Figure 5.2a–d highlights the relative changes in the consumer isocline
as a parameter is varied. Any of the above parameters drive this geometric
sequence. For simplicity let us consider the maximum attack rate in the R-M
model, a_{max}, and assume we are increasing it. Figure 5.2a displays the isoclines for low levels of a_{max}. In this case, low attack rates produce a small
consumer density relative to the resource density. The consumer is quite close
to the feasibility point, where $C = 0$, simply because it is a poor consumer.
The actual feasibility point occurs when the vertical consumer isocline lies on
top of the carrying capacity, K. In figure 5.2e, region I displays the corresponding eigenvalue response for low attack rates. Near the "feasibility" point a low
a_{max} produces monotonic C-R dynamics. The consumer isocline in figure 5.2a
is in the monotonic zone because it is to the right of the rightmost dashed line.
Recall from chapter 4 that all increases in population growth rates inspire more
rapid return times and greater stability when the dynamics are monotonic or
nonexcitable. The descending part of figure 5.2e, which shows an increase in

stability, shows that increases in coupling strength, a_{max}, stabilize the dynamics when they are in the monotonic region I, as expected from chapter 4.

As the parameter is further increased, the consumer isocline shifts further left relative to the resource isocline (figure 5.2b and c). Eventually, the consumer isocline passes through the dashed line into region II as the eigenvalue becomes complex (figure 5.2e). The dynamics in region II are now all characterized by oscillatory decay, and increases in the parameter a_{max} inspire greater oscillatory decay or overshoot population dynamics. The dynamics have become excitable. Thus, at the moment the consumer isocline pushes through the transition to oscillatory decay into region II, the parameter switches from having a stabilzing role to having a destabilizing role. Again, this result is consistent with the single-population level results in chapter 4.

For this reason, the point of transition to oscillatory decay marks the maximum stability (figure 5.2e). Still further increases in the parameter a_{max} push the consumer isocline through the transition to sustained oscillations (figure 5.2d; called a Hopf bifurcation), and the dynamics become so unstable as to produce cyclic fluctuations (region III, figure 5.2e).

We have just walked through one parameter (a_{max}) and shown that, as it increases, it first briefly stabilizes the dynamics before destabilizing the dynamics over a large region of parameter space. Biologically, this result says that if the C-R dynamics show overshoot potential (i.e., are excitable), increases in the flux through the C-R interaction tend to destabilize the dynamics. This same result also occurs for all other parameters except r; however, increasing some parameters (a_{max}, e, K) moves us from left to right in figure 5.2a–d, while increasing others (R_o, m) moves us from right to left in figure 5.2.

Increasing K, a_{max}, or e increases the relative coupling strength and so moves the consumer isocline from left to right (figure 5.2a–e), while one must decrease the parameters R_o and m in order to increase the relative coupling strength and move the isoclines from left to right in figure 5.2a–d. Recognize that decreasing the half-saturation parameter, R_o, increases the consumption rate and that this biologically drives a stronger coupling of consumer and resource. Additionally, reducing consumer mortality, m, reduces the amount of energy lost by the consumer and so effectively increases the relative coupling strength (coupling terms:loss terms). Thus, all the above parameters can be modified (increased or decreased) to produce the result indicated in figure 5.2e. From a theoretical perspective this gives a beautiful general result with a consistent interpretation in terms of energetics. High relative fluxes through a consumer-resource interaction tend to ultimately be destabilizing.

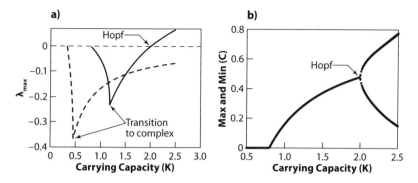

FIGURE 5.3. Local stability and dynamic response of consumer-resource interactions to increasing K. (a) Solid curve is the R-M model, which crosses into
cyclic dynamics (marked by the Hopf designation). The dashed curve is the
Lotka-Volterra with logistic growth. The local stability response is qualitatively
similar except it does not cross the Hopf, as is well known. (b) The maxima and
minima for the R-M model, which show the same destabilzation effect of K.
Parameters for the R-M: $r = 1.0$; $a_{max} = 1.50$; $R_o = 0.40$; $e = 0.50$; $m = 0.50$.
Parameters for the L-V: $r = 1.0$; $a = 1.50$; $e = 1.0$; $m = 0.50$.

Figure 5.3 again shows a specific example of the above sequence of events
for the R-M model, where K is increased in order to mimic increasing the
resource standing crop. Figure 5.3a shows that this increase in K first stabilizes
the dynamics briefly before it is dominated by a large region of destabilization
(solid curve). This result in the R-M model is the familar "paradox of enrichment" (Rosenzweig, 1971). Recall from the box flow diagram in figure 5.1
that K governs the amount of energy lost from the system. As K increases, for
example, the loss of energy (more specifically damping in this case) is weakened, and this increases the relative strength of the flux of biomass through the
C-R interaction. Thus, increases in K should change the shape of the isoclines
as in figure 5.2a–d, and it does.

Also shown in figure 5.3a is the behavior of the Lotka-Volterra model with
logistic growth (dashed curve). The dynamical response of the L-V model is
qualitatively similar to that of the R-M model but never reaches sustained
oscillations. The L-V model with logistic growth does not follow the traditional phrasing of the paradox of enrichment in that it is not necessarily driven
to wild oscillations by increased K. Importantly, though, it still shows the same
dominant destabilization phase. Again this destabilization phase coincides
with the onset of excitable dynamics (figure 5.3a).

In this book I prefer to think of the paradox of enrichment less based on
its extreme statement—increased production leads to extinction—and more

based on the fact that C-R models generally find that increased biomass flux through the C-R interaction tends to destabilize the interaction. This less contentious argument will prove to be powerful in allowing us to understand the implications of food web structure.

Here, I have also shown that this local destabilization has similar nonequilibrium consequences for stability as K is increased (figure 5.3b). Specifically, as K increases, the system goes from stable, with no variance, to cycles that tend to increase in amplitude with increasing K (figure 5.3b). All else equal, large amplitude cycles approach low population densities and so are at increased risk of extinction (e.g., a bad winter at low densities may extinguish such a population).

While not mentioned in the "paradox of enrichment" story, I have shown here that a suite of other parameters (a_{max}, e, R_o, m) have the same consequences as changing K. Varying each of these parameters operates to vary the strength of the coupling term relative to the loss terms (figure 5.1). The changing relative position of the isoclines is a geometric way of stating this changing biology. It turns out that the paradox of enrichment could just as easily have been called "the paradox of coupling strength" because any term that operates to increase the relative flux through the interaction ultimately makes the system less stable. This is a far more general result than the paradox of enrichment.

To this point I have emphasized the results in terms of the relative strength of the flows identified in the mass balance diagram in figure 5.1. It remains to relate this result to classic definitions of interaction strength. It turns out that any parameter that moves the relative position of the isoclines as discussed in figure 5.2a–d has a clear influence on May's interaction strengths. Specifically, such a changing geometry mathematically means that the interspecific interaction strengths (off-diagonals in a Jacobian matrix) tend to gain strength relative to the intraspecific interaction strengths (diagonal terms). This is a nice result because it suggests that the intuitive biological mechanism of relative flux rates maps cleanly to our traditional measures of interaction strength. This allows us to state the above general result in terms of both coupling and interaction strength, which I will do below.

There is another important biological outcome of increasing the coupling terms relative to the loss terms. This increase in flux also tends to increase the C:R biomass ratio. A number of people have pointed out that higher C:R ratios tend to correlate with less stability (Raffaelli, 2002; Neutel et al., 2002; Murdoch et al., 2003), and this pattern evidently holds here for the R-M model. This is not surprising in light of the generalization contributed here: Given that the relative amount of biomass flux increases through an interaction, one expects that the consumer density will rise at the expense of the resource

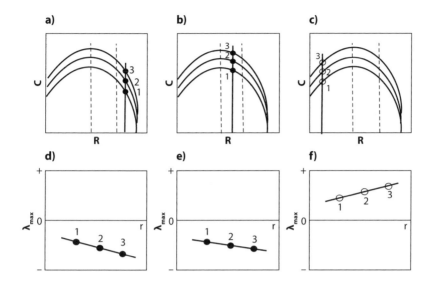

FIGURE 5.4. Qualitative response of consumer-resource isoclines to variation in the parameter, r, in the R-M model. Specifically, increasing r drives a change in the resource isocline that matches (a) – (c) and produces the eigenvalue, λ, response shown in figure 5.2d – f.

density. Because of this, we expect a strong correlation between the biomass pyramid and the stability such that Eltonian biomass pyramids (i.e., where the resource biomass is far greater than the consumer biomass or the C:R ratio is low) tend to be more stable than inverted biomass pyramids (i.e., where the consumer biomass is greater than the resource biomass or the C:R ratio is high). This is a somewhat intuitive result, as it suggests that the more top-heavy a biomass pyramid, the less stable the ecosystem. Biomass pyramids, therefore, may be an indicator of stability.

5.2.2   REMOVING EXCITABILITY IN C-R INTERACTIONS: THE INFLUENCE OF RESOURCE UPTAKE, r

Now let us turn to the case of increasing r (figure 5.4a–f). The parameter, r, has no direct influence on the coupling terms. The geometric response of the isoclines to increasing r shows this because the consumer isocline does not shift left relative to the resource isocline (compare figure 5.4 to figure 5.2). Two important cases exist illustrating variation in the parameter r. First, when all parameters are chosen such that the consumer isocline intersects the stable half of the resource isocline (i.e., on the right-hand side of the hump in the resource isocline), increasing r does the exact opposite of all our previous results—it destabilizes when the eigenvalues are real and stabilizes when the eigenvalues are complex. Second, when the parameters are chosen such that the interaction is unstable, or the consumer isocline lies on the left side of the hump, increasing r increases the maximum eigenvalue (makes it more positive).

These results make sense in light of how increasing r changes the isocline shapes. Increasing r tends to increase the magnitude of the negative slope of the R isocline when it intersects the C isocline on the stable side (right-hand side) of the hump in the resource isocline. This makes the diagonal R term in the interaction strength matrix more negative, which has a stabilizing tendency. Increasing r also makes the off-diagonal coupling term weaker. Both of these changes imply that the resource dynamics dominate the local C-R dynamics—in a sense they are making the system less complex as we increase r.

In the terminology developed above, increasing r reduces the strength of the interspecific interaction strength *relative* to the intraspecific interaction strength. On the unstable, or left-hand, side of the hump in the resource isocline, increased r increases the positive slope of the R isocline, which means that the diagonal term is more positive, while the off-diagonal coupling term is still weakened. Now, r is inspiring the unstable equilibrium to depart even faster from the equilibrium, although with less spin on the departure (since the off-diagonal term is weakened, it makes this departure less oscillatory).

Putting this altogether, increasing r makes the resource dynamics strong relative to the consumer dynamics because increases in r manifest largely in the resource uptake term (i.e., rR in figure 5.1) and the loss term ($-rR^2/K$ in figure 5.1). The result is that r tends to stabilize dynamics when they are already stable but have oscillatory decay (DeAngelis and Goldstein, 1978). Specifically, r weakens the relative strength of the interspecific interaction strengths by increasing the loss term (note that on the right-hand side of the resource isocline R^* is close to K, so the loss term, $-rR^{*2}/K$, is large and negative in this case). Here, increasing r shoots the solution back more directly to the equilibrium, removing some of the overshoot dynamics (coupling dynamics).

When the solutions are unstable (repelling from an attractor), the linear term, r, causes the equilibrium to only repel more, and so increasing r makes the system that much more unstable [in this case the consumer isocline is on the left-hand side of the resource isocline where the loss term is weak (R^*/K is small), and so the positive rR term dominates the diagonal]. The increase in r therefore makes an unstable equilibrium (positive real eigenvalue) more unstable (a more positive real eigenvalue).

It should be pointed out that a discrete version of this model can be expected to give a different answer than the one above. In chapter 4 we found that increases in r decreased stability once the eigenvalues became excitable.

In continuous-time models, this excitability or overshoot potential is never realized. We will find in the next chapter, which explores the influence of biological lags, that in fact once r is associated with a lag, it acts in a way that is entirely analogous to increasing the carrying capacity, K. That is, increases in production via r first stabilize before destabilizing C-R interactions once the system experiences overshoot dynamics (recall from chapter 4 that a large r drives population oscillations or cohort cycles). If lags exist, r may behave dynamically like all other parameters—increased flux into the basal species will ultimately tend to destabilize.

5.3   SUMMARY

The above results can be biologically stated from a flux-based perspective (figure 5.1) as well as from an interaction strength perpective as follows:

Conjecture 5.1:
(i) Increasing the coupling strength between a consumer and a resource, relative to the strength of the loss terms, tends to destabilize an already excitable C-R interaction. Simlarly stated with interaction strengths;

(ii) Increasing the interspecific interaction strength between a consumer and a resource, relative to the strength of the negative intraspecific interaction strengths, tends to destabilize an already excitable C-R interaction.

These conjectures give cause for some further reflection. It is informative, for example, to think about the opposite of the above statement, and doing so produces the following corollaries.

Corollary 5.2:
(i) Any biological mechanism that operates to reduce the strength of the coupling in a C-R interaction, relative to the strength of the loss terms, tends to stabilize an already excitable consumer-resource interaction.

(ii) Any biological mechanism that operates to reduce the strength of the interspecific interaction in a C-R interaction, relative to the strength of the negative intraspecific interaction stengths, tends to stabilize an already excitable consumer-resource interaction.

Let us consider how this conjecture can be used with some examples. Take a lynx-hare interaction that naturally has strong interspecific interaction strengths; the dynamics are excitable in that they are either of the oscillatory decay kind (quasiperiodic) or actually cycling. Suppose now that a new species, say an introduced cougar, is added to the mix such that it is stable and it consumes this same resource (hare) but does so only weakly. This added consumption by the cougar reduces the resource production available to the lynx (i.e., reduces the K available to the lynx). We would expect from the corollary above that in this hypothetical case the lynx-hare dynamics would become more stabilized with the addition of the cougar. Similarly, nontrophic interactions can have the same outcome. Let us take the case that, in the presence of an additional predator, the lynx reduces its foraging time and instead moves away from the competing predator. This is behavior that reduces consumption rates, ultimately muting flux rates and stabilizing the excitable interaction. We will return to these more formally in the not so distant future as they begin to form the simple mathematics for understanding the dynamical implications of adding subsystems together, a mathematical approach that is required if we are to extend from consumer-resource interactions to food web module approaches and finally to whole systems.

5.4 FURTHER ASSUMPTIONS ABOUT THE C-R MODEL

To this point we have considered a well-founded consumer-resource model, but other attributes of C-R interactions have been sidestepped. Here, we briefly discuss some of these common attributes with respect to the conjecture and corollary above.

5.4.1 SELF-LIMITATION

In the Rozenzweig-MacArthur C-R model, self-limitation occurs only in the resource, but a consumer may be similarly afflicted such that it experiences density-dependent mortality (e.g., interference), which manifests as a stronger loss term, $-mC^2$, in the consumer growth function. In terms of isoclines, this self-damping term drives the C isocline from a straight line into a curved isocline. The geometry behind this additional biology therefore means that this type of C-R intersection more frequently intersects the right-hand side of the R isocline relative to the straight isocline of the R-M model. Further, it obviously means that the consumer is much less efficient in its use of biomass consumption because more is lost from the system (i.e., there are increased loss terms). Therefore, we immediately know that this is stabilizing (Tanner, 1975). Aside from this relative increase in stability, this model still qualitatively responds in a similar fashion to increased interspecific interaction

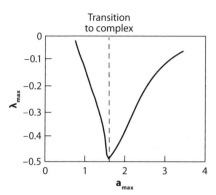

FIGURE 5.5. Local stability response of an R-M model with a consumer damping term to an increasing attack rate a_{max}. The qualitative result of stabilization followed by destabilization once the dynamics are excitable holds. This model delays the transition to oscillatory decay (i.e., overshoot dynamics) longer than the classic R-M model, but otherwise qualitatively similar results occur. Parameters: $r = 1.0$; $K = 1.75$; $R_o = 0.40$; $e = 0.50$; $m = 1.50$.

strengths. As an example, figure 5.5 shows the response of this model to an increase in attack rate. Here, as before, we see the familar stabilization phase give way to a destabilizing phase when excitable dynamics are induced. Under this model we would expect the world to be far less oscillatory (production would more often tend to increase stability than to decrease stability), but otherwise the responses to interspecific interaction strength are the same.

5.4.2 MODIFIED FUNCTIONAL RESPONSES

There are numerous variants of the functional response, but all reasonable ones that I am aware of produce the same qualitative results as the C-R model. The ratio-dependent model (the functional response depends on the ratio of resource density to consumer density) produces an isocline geometry not far from that of the self-damping case above (Cantrell and Cosner, 2001; Rall et al., 2008). This may not be surprising because ratio dependence in the functional response is a way of assuming interference between consumers during consumption (Yodzis, 1994). Thus, the model is more stable than the classic type II functional response.

The type III functional response, which assumes reduced consumption rates at low resource densities, tests the above conjectures a little more because an increase in interspecific interaction strength, again through a_{max}, drives a stabilization phase, a destabilization phase, and then another stabilization

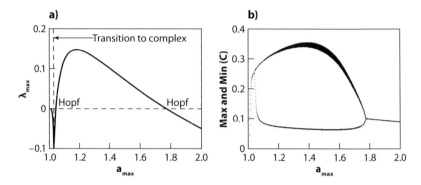

FIGURE 5.6. (a) Local stability response of an R-M model with a type III functional response to increasing attack rate a_{max}. The qualitative result of stabilization followed by destabilization once the dynamics are excitable holds; however, here the dynamics eventually become stabilized again as the consumer eventually consumes less per unit individual at low resource densities and so effectively decouples the consumer dynamics from the resource dynamics. See text for discussion. (b) Dynamical response over same region showing that the oscillations increase and then decrease again. Parameters: $r = 1.0$; $K = 1.0$; $b = 0.01$; $e = 0.50$; $m = 0.50$.

phase (figure 5.6a)—all correspond to the predicted transitions between purely real (monotonic) and complex eigenvalues that beget overshoot dynamics and eventually sustained oscillations (figure 5.6b).

As mentioned, under the assumption of a type III functional response, at low R densities consumers have trouble feeding or finding resources, and so the resources are little influenced by consumers at low resource densities—the added biology of the type III response implies that the coupling term is weak at low R densities. This means the system should be made less excitable in this region, and this is what we find; sustained oscillations are lost, and the dynamics strongly attract locally. The results resonate with the conjecture and corollary above, as the added biology weakens the coupling strength at low resource densities.

Finally, it is well known that donor control is stabilizing (DeAngelis, 1975). This, too, makes sense in light of the corollary. Donor control means the negative interspecific interaction strength term for the resource is effectively zero. Thus, donor control decouples the resource from the consumer or, stated differently, donor control removes the excitability of the consumer-resource interaction dynamics.

5.5   SOME NONEQUILIBRIUM THOUGHTS

While we have thought about how the relative interaction strengths influence a C-R interaction near an equilibrium, it remains a possibility that nature has developed attributes that may act to stabilize or destabilize nonequilibrium dynamics. Consider a C-R oscillation such that there are times when the consumer is in the act of overshooting the resource. That is, the consumer is at high densities, and the resource has begun to be suppresssed to low densities. Any density-dependent mechanism that operates to mute flux rates between these two species at this point clearly will act to reduce overshoot and stabilize. Further, consider the part of the opposite end of the C-R cycle, where C is low and R is high. Here, increasing flux between these two species increases C's growth rate when needed (i.e., low density) and starts to reduce R when it is high. If any density-dependent mechanism operates such that it tends to increase and decreases flux rates at the appropriate time, we expect it to be a potent stabilizing mechanism for fluctuating populations.

Spatial heterogeneity occasionally creates scenarios where such density dependence occurs (Crowley, 1981; McCauley et al., 1993, 1996). Imagine a consumer feeding on a patchy resource. At the local scale a consumer may find itself at high densities and soon in a patch of low R. Given heterogeneity, it becomes compelling for a mobile consumer to behave by finding a better patch (i.e., high R, low C). These are the precise conditions for nonequilibrium stabilization discussed above, and this result is well known in the population ecology literature [e.g., McCauley et al. (1996)]. Additionally, one can imagine the resource moving more than the consumer. Similarly, any kind of meaningful avoidance behavior will readily generate the same conditions. Given that both move at similar scales, this type of density-dependent stabilization is essentially lost (McCauley et al., 1993). We will return to concrete examples of this as we move into more complex community dynamics.

Early models and experiments wrestled with nonequilibrium stabilization. For example, Nicholson and Bailey (1935) discussed how such responses could act to stabilize predator-prey interactions when they found that their now famous model produced runaway oscillations. In response to the runaway growth of the model, they argued that the "retarding influence of such factors as scarcity of food or suitable places in which to live is bound to be felt." Similarly, Huffaker (1958) set up spatially heterogeneous mite experiments in order to curb the excessive runaway growth and extinction of predatory mites. In the heterogeneous case, the experimental setup allowed the prey to stay one step ahead of the predatory mites at key times (low densities), muting predator-prey consumption rates and fostering persistent but oscillating population dynamics.

5.6   C-R DYNAMICS IN NATURE

Some old and recent microcosm approaches have isolated natural C-R interactions in a bottle and then varied a key biological parameter (e.g., productivity and attack rate). These studies have consistently found results that agree with C-R theory. Luckinbill (1973), for example, modified attack rates in a predator-prey aquatic microcosm by adding methylcellulose, which reduced the attack frequency of the predator. The above theory predicts that such an experiment would tend to stabilize the interaction, and the experiment of Luckinbill (1973) indeed found prolonged coexistence of the otherwise very unstable interaction. As mentioned in chapter 2, Fussmann et al. (2000) varied production and experimentally reproduced the paradox of enrichment. In fact, their experiment even showed evidence that the natural system crossed the Hopf boundary, or transition to cyclic dynamics. The experiment by Fussmann et al. (2000) actually provides a further test of the above theory in that their experimental manipulation had two consequences:

(i)   Low to moderate dilution rates enriched the system (i.e., increased K), and

(ii)   High dilution rates produced a sudden striking increase in mortality (loss terms) that outweighed any production increases. At high dilution rates organisms were litterally swept out of the chemostat.

The above theory suggests, then, that we would first expect destabilization with increased enrichment followed by stabilization with increased loss. Their experiment, in fact, found exactly this (see figure 2.11).

There are also ways to consider the results of this theory within a whole ecosystem context. As an example, there has been much recent work on contrasting aquatic and terrestrial ecosystems (Cyr and Pace, 1993; Hairston and Hairston, 1993; Shurin et al., 2002; Cebrian et al., 2009). The results from this empirical research make sense in light of the general theory of C-R interactions.

Aquatic ecosystems, on average, tend to have smaller-bodied organisms at each trophic level (Rooney et al., 2008). Some of this size difference likely manifests from the need for the basal pelagic trophic level (i.e., algae) to remain relatively bouyant in order to harvest energy from the sun—rapid sedimentation would clearly be a significant disadvantage in aquatic ecosystems (Hairston and Hairston, 1993; Shurin et al., 2002). These smaller organisms, all else equal, have higher per capita attack rates and death rates, meaning that the aquatic system is characterized by a more rapid turnover than terrestrial ecosystems (Hairston and Hairston, 1993). Further, aquatic organisms tend to also have body size ratios between predator and prey that promote higher consumption rates (Shurin et al., 2002). Finally, the stoichiometric similarity between consumers and resources is more aligned, suggesting that material transfer should be more efficient in aquatic ecosystems (Shurin et al., 2002).

It is also the case that aquatic ecosystems have organisms that tend to carry a more edible biomass. Terrestrial plants are often composed of a large portion of relatively inedible lignins that form the plant's core biomass. Additionally, there is a tendency for the skeletal structure (i.e., bones) of larger animals to increase with size and metabolic type. Thus, large terrestrial mammals have a skeletal structure that is significantly larger per unit of biomass and less edible than that of their aquatic high trophic level counterparts, fish.

Taken together, these observations suggest that rates of flux between consumer and resource (i.e., the coupling strength) ought to be much higher in aquatic than in terrestrial systems on average. Cyr and Pace (1993) noticed this pattern empirically for herbivory. The existing evidence also suggests that the higher coupling strengths in aquatic systems likely outweigh the corresponding increase in loss terms (i.e., the rapid loss of smaller organisms in aquatic ecosystems). Yodzis and Innes (1992) scaled and parameterized a C-R model such that one of the key parameters was ingestion rate relative to metabolic cost. This parameter, in a sense, then weighs the coupling strength relative to the loss (called "ecological scope" by Yodzis and Innes (1992)]. They found that vertebrate ectotherms (e.g., fish) tended to have a much higher scope than vertebrate endotherms (e.g., terrestrial mammals). Further, allometric reasoning suggests that as body size increases, metabolic costs should cancel out increased consumption rates; however, given that aquatic communities also have prey that are more easily converted into biomass, this suggests that smaller organisms ought to have a higher coupling strength relative to the loss terms. Thus, one would expect aquatic communities to have higher interspecific interaction strengths relative to intraspecific interaction strengths than terrestrial communities—the ratio of nondiagonals to diagonals in the interaction matrix should be larger than in the terrestrial case.

We know, then, from the above development that this suggests that aquatic consumer isoclines, on average, should tend to lie further left relative to the resource isocline than for terrestrial organisms. The predictions that follow are

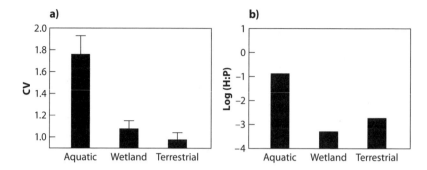

FIGURE 5.7. Cross-ecosystem differences in (a) the coefficient of variation and (b) the ratio of consumers to resources, measured as [log herbivore (H) biomass] : [log primary producer (P) biomass]. (b) Reproduced with permission from Cebrian et al. (2009). Note that wetland systems were considered terrestrial in Cebrian et al. (2009) but are isolated here for comparison to other figures.

(1) The C:R ratio should be higher in aquatic organisms than in terrestrial organisms, and

(2) population dynamics in aquatic systems should be less stable (more variable) than in terrestrial ecosystems.

Consistent with this, aquatic systems are known to occasionally show inverted biomass pyramids (consumer-heavy or non-Eltonian biomass pyramids). Shurin, Cebrian, and others have rigourously attacked this issue using meta-analysis to show that there is a distinct tendency for aquatic ecosystems to have a higher C:R ratio (Shurin et al. 2002; Cebrian et al. 2009) Figure 5.7b shows the meta-analysis of C:R ratios, here herbivore:plant ratios, from Cebrian et al. (2009). The data are categorized into three ecosystem types (terrestrial, aquatic, and wetland). The results suggest that terrestrial ratios are more resource-heavy, as predicted.

On the other hand, little has been done to consider the stability implications of the higher relative coupling strengths of aquatic ecosystems. To explore this, Rip and McCann (forthcoming) collected 500 time series from the global population database. They then separated the time series by trophic position and ecosystem type (aquatic, terrestrial, and wetland). As predicted, figure 5.7a shows that the variance in population dynamics in these systems is much greater in aquatic ecosystems than in terrestrial ecosystems. These results therefore agree with the general theory of C-R interactions because aquatic systems tend to have both fewer Eltonian pyramids and more variable, or less stable, dynamics. Curiously, the intermediately positioned ecosystems,

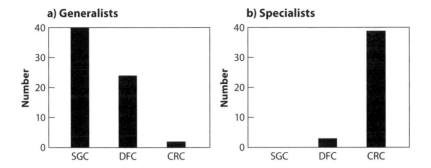

FIGURE 5.8. Distribution of cycles among classes defined by scaled period. SGC, Single-generation cycles; DFC, delayed-feedback cycles; CRC, consumer-resource cycles. SGC and DCF are forms of single-species cycles. (a) Generalists. (b) Specialists. Generalists are dominated by cycles of the single-species variety, while specialists tend to be of the consumer-resource variety. Figure modified from Murdoch et al. (2002).

wetlands, which have intimately linked terrestrial and aquatic components, also show intermediate levels of stability.

Along similar lines, trade-offs suggest that specialists should have higher attack rates on any individual resource than generalists. Numerous researchers have argued that the performance of a given consumer on a distribution of resources can be expected to be relatively invariant across species [e.g., MacArthur and Pianka (1966); Levins (1968)]. Given this, specialists by definition have a narrower performance spectrum with a much increased maximum performance relative to generalists. Empirical data agree with this prediction [e.g., Strickler (1979); Drummund (1983)]. The results from consumer-resource theory suggest that this heightened consumptive ability ought to mean that specialists are less stable, or more variable, than generalists. A number of researchers have followed up on this idea tangentially. For example, in his book on complex population dynamics, Turchin et al. (2003) found that, generally speaking, specialist predators were responsible for complex dynamical phenomena. In another contribution, Murdoch et al. (2002) found evidence that most longer-period consumer-resource cycles were prominent in specialist predators, while generalist predators, decoupled from any specific prey, tended to show evidence of shorter-period, single-species fluctuations (figure 5.8a and b). The split between these two types of cycles is dramatic (figure 5.8). This is a fascinating example of an approach that begins to ask how food web properties influence population dynamics.

There remains the well-known example of Hanski et al. (1991) that examines lemming population dynamics across a gradient in the relative percentage of generalist predators. Again, one would expect that the dynamics would be more oscillatory and variable toward the zone dominated by specialists, and this is what happens. As one moves south, the percentage of generalist predators increases, and simultaneously the variability of the lemmings is reduced. Lemmings are highly reproductive and can attain enormous densities—this coupled with high attack rates suggests that they are perfect candidates for oscillations. The role of specialist predators in the lemming cycle, though, is not definitive, and further work has pointed out that herbivore-plant or maternal effects may also play a significant role in producing this range in dynamics, although predation appears most likely given the existing evidence (Turchin and Hanski, 2001).

5.7   SUMMARY

1. Robert May's interaction strength (a Jacobian matrix) can be understood in terms of important flux rates between species (coupling strength) and loss rates from a single species (e.g., consumer and resource loss rates).

2. **The rule of relative coupling strength**: Increasing the coupling strength between a consumer and a resource, relative to the strength of the loss term, tends to destabilize an already excitable C-R interaction.

Stated similarly, increasing the interspecific interaction strength between a consumer and a resource, relative to the strength of the intraspecific interaction strengths, tends to destabilize an already excitable C-R interaction.

3. **The coupling strength corollary**: Any biological mechanism that operates to reduce the strength of the coupling in a C-R interaction, relative to the strength of the loss terms, tends to stabilize an already excitable consumer-resource interaction.

Stated similarly, any biological mechanism that operates to reduce the strength of the interspecific interaction in a C-R interaction, relative to the strength of the intraspecific interaction stengths, tends to stabilize an already excitable consumer-resource interaction.

4. Theoretical examples from the literature agree with (3) above. For example, self-limitation, donor control, and ratio dependence all would be expected to be stabilizing given (3).

5. Microcosm experiments and empirical data are presented and show consistency with the general C-R theory presented here.

CHAPTER SIX

Lagged Consumer-Resource Dynamics

There are a number of reasons why the biology of organisms creates lagged effects on populations dynamics. Because of this, the unstructured continuous consumer-resource model in chapter 5 can be argued to miss some of the important dynamical influences of realistic biological lags. Lags occur, for example, when a population reproduces at regular and synchronized intervals in time. Similarly, organisms are almost always comprised of various life stages that often occupy different habitats and feed on different prey items. This stage structure introduces lags into the dynamics of an adult class, as exemplified when a large cohort of juveniles matures into the adult class. Such stage structure also implicitly couples the dynamics of one stage and possibly one habitat with another stage and another habitat. Many organisms show such behavior, and there are strong arguments that discrete equations better model regular but discrete aspects of organismal life history (Hassell, 1978). It behooves us, therefore, to consider the implications of this ubiquitous population level structure and place these results within the framework defined in the previous chapters. We will find that this additional structure has some implications, and fortunately a number of researchers have made significant progress in this area [see Murdoch et al. (2003) for a review].

Here, I will concentrate on how C-R cycles, or the lack thereof, interact with population level dynamical phenomena. In the sense discussed above, both the underlying population level and the C-R interaction can produce cycles independent of each other, and so it becomes necessary to consider how these two underlying subsystems interact. In some cases, one subsystem may excite the other into complex dynamics, while in other cases, one subsystem may mute an otherwise unstable interaction. In this brief and preliminary chapter, I will argue that weak and inherently stable C-R interactions can mute a potentially unstable population level phenomenon. Additionally, a dynamically decoupled stable stage class can strongly stabilize other stages and the C-R interaction. In what follows, I will first consider discrete consumer-resource interactions and then turn to a brief consideration of stage-structured consumer-resource interactions.

6.1 DISCRETE CONSUMER-RESOURCE INTERACTIONS

Let us start with the following simple discrete analog of the consumer-resource interaction with logistic growth analyzed by Neubert and Kot (1992):

$$
\begin{cases}
R_{t+1} = R_t + rR_t(1 - R_t/K) - aR_tC_t, \\
C_{t+1} = eaR_tC_t,
\end{cases}
\tag{6.1}
$$

where r is the per capita rate of increase in the resource, R, K is the carrying capacity of the resource, a is the attack rate of the consumer, C, on R, and e is the conversion efficiency of resource biomass into consumer biomass.

This model produces a rich array of dynamics. System (6.1), for example, can have a C-R-driven cycle (see chapter 5) as well as cycles imposed by the lags inherent in the discrete system (see chapter 4). Whenever multiple oscillators exist, they can interact. In such cases, dynamical systems readily produce complex dynamical phenomena including chaotic dynamics, and the above discrete C-R model is no exception (Neubert and Kot, 1992).

In the continuous C-R model in chapter 5, increasing resource growth rate, r, did not make a stable system less stable. In fact, it did the opposite. This occurred despite the fact that increasing carrying capacity, K, in the continuous C-R model destabilizes dynamics with increased production. Thus, one aspect of increasing production (r) stabilized dynamics, and another aspect of production (K) destabilized dynamics. However, with discrete equations we already know that r inspires nonequilibrium resource dynamics in the discrete logistic. Because of this, it seems likely that r acts potentially more like K in the discrete C-R model.

Figure 6.1 shows the response of the maximum real eigenvalue as the per capita resource growth rate, r, is increased. In this case, I chose parameters for the underlying C-R interaction that produced stable dynamics (i.e., weak attack rates relative to loss rates). In such a case, increasing r first further stabilizes the whole system, reaching maximum stability when the eigenvalue is zero. After that the eigenvalue becomes negative. Recall that negative eigenvalues for discrete systems imply excitable dynamics. Thus, we expect the system to be destabilized with increasing production, r, as occurs for K in the continuous model in chapter 5. This is exactly what happens, as the eigenvalue grows in magnitude. Further increases in production now drive greater overshoot dynamics and slower return times. With a high enough r the system permanently fluctuates (i.e., when $|\lambda| > 1$; figure 6.1). Unlike the influence of K, which inspires C-R cycles, increasing r excites the underlying resource dynamics, R, to cycles and more complex dynamical phenomena. Nonetheless, with

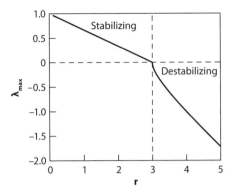

FIGURE 6.1. Local stability response of a discrete L-V model with logistic to inceasing r. The qualitative result of stabilization followed by destabilization once the dynamics are excitable (i.e., a negative eigenvalue for discrete dynamics) holds as it does for the continuous models presented in chapter 5. See the text for a discussion. Other parameters: $K = 1.0$; $a = 1.50$; $e = 1.0$.

discrete lags C-R theory now consistently argues that production increases ultimately tend to be destabilizing.

The problem of understanding the dynamics in this discrete C-R system is clearly a more complex problem than in the continuous C-R case in chapter 5 precisely because the problem now needs to consider multiple underlying oscillators. The interaction of potentially oscillatory subsystems is something that will arise frequently when we move to food web dynamics where interacting subsystems with oscillatory potentials abound. While it is well known that interacting cycles can lead to complex dynamics that produce many local minima and maxima, like chaos, it is less appreciated, and equally important, that two interacting subsystems can inhibit potentially cyclic dynamics. For a full ecological discussion of chaos, see Hastings et al. (1993). The relative interaction strength conjecture and corollary in chapter 5 suggest that this can frequently be the case. Specifically, if a relatively weak and inherently stable subsystem removes flux from the other subsystems, it can effectively act to increase dissipative loss from the potentially oscillating interaction and so stabilize the dynamics.

As an example of this inhibition of underlying cyclic dynamics, let us again look at the above model, but this time let us start with a relatively modest C-R interaction strength (i.e., the interaction is weakly cyclic). As before, the model experiment will increase r and monitor the dynamical response of the consumer and the resource. In figure 6.2, I display the dynamics for both the C-R model response to increasing r (gray dots) as well as the response of

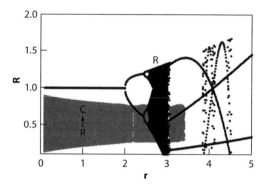

FIGURE 6.2. Dynamics on the attractor of the discrete C-R model over a range of *r* (gray points) and underlying resource dynamics (without a consumer) for the same *r* values (black points). Results show a weakly cyclic underlying C-R interaction (gray points) relative to the resource dynamics alone, R, (black points) until suddenly the C-R interaction is totally destabilized and the system crashes. This crash occurs at approximately $r = 3.50$, at which point the gray data stop indicating the extinction of the system. Other parameters: $K = 1.0$; $e = 1.0$; $a = 2.20$.

the discrete logistic resource model (black dots). I include the logistic resource dynamics in order to compare how the whole C-R system (6.1) responds relative to the underlying R-subsystem dynamics alone. One can see that the discrete logistic model (black dots) undergoes the regular period doubling to chaos with increases in *r*. Meanwhile, the C-R dynamics (gray dots) remain well-bounded (i.e., do not fluctuate wildly) until $r > 3.5$, at which point they become wildly unstable and crash. For a full account of the complex dynamical possibilities of this model, see Neubert and Kot (1992).

The discrete C-R system therefore experiences much more bounded population dynamics, from around $r = 2.5$ to $r = 3.5$, than the population dynamics of the resource system alone. In essence, the consumer acts to stop the resources from attaining high resource densities. In doing so, the consumer prevents the resource, R, from reaching high densities where overcompensatory responses drive population level cycles. As long as the C-R dynamics are stable or modestly cyclic as in figure 6.2 (i.e., the cycles are not so large as to allow R to reach overcompensatory densities), this result remains. In a sense, then, weak to intermediate-strength C-R interactions can stabilize this system, and this is something we will see again and again in coming chapters. The above stabilization results are robust to reasonable variants on the discrete C-R interaction model.

Following the results of Murdoch et al. (2002), this stabilizing influence of the consumer can be especially important in the case of generalist

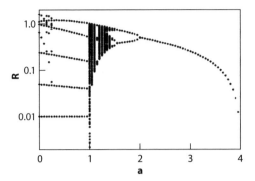

FIGURE 6.3. Consistent with Murdoch et al. (2002), model system (6.2) C-R dynamics are stablized by generalists with increased attack rates. Resource dynamics undergo a period-doubling reversal to stable equilibrium population dynamics. At high enough attack rates, though, generalists can clearly suppress resources to extinction, albeit always obeying equilibrium dynamics. Parameters: $r = 4.0; K = 1.0; C = 1.0$.

consumers. In an empirical survey of the causes behind population cycles, Murdoch et al. (2002) found that most cases of cyclic dynamics due to population level lags—as opposed to those cycles produced by consumer-resource interactions—occurred on prey that had generalist predators. They argued cogently that the generalist consumer is effectively decoupled from the resource, leaving cyclic phenomena to be driven solely by population level feedbacks. Nonetheless, the presence and strength of the generalists—even if decoupled—can act to limit the density of R as well as the magnitude of cyclic dynamics. In such a situation, removing resource biomass acts as a constant increase in dissipative loss. Thus, if our arguments from chapter 5 are correct, we expect such an increase in dissipative loss to inhibit the population cycles of the resource either by reducing the amplitude or by driving the population dynamics to a new lower stable equilibrium.

Let us consider this more explicitly by looking at a discrete resource model with a constant number of generalist consumers:

$$R_{t+1} = R_t + rR_t(1 - R_t/K) - aCR, \tag{6.2}$$

where a is the attack rate and C is the constant amount of generalist consumers that are dynamically decoupled from the resource dynamics. Varying the attack rate in this condition acts to potentially stabilize the dynamics through what is known as a period-doubling reversal (figure 6.3). The decoupled generalist is an extremely potent stabilizing force. Note that because of the decoupling of the consumer dynamics from the resource dynamics, even relatively high interspecific interaction strengths on R by C can still stabilize the variance.

Clearly though, if a or C is high enough, the interaction may be destabilized in the most complete sense in that the resource equilibrium disappears (i.e., generalist C suppresses R to local extinction). This in fact occurs at an attack rate of about 4.0 (figure 6.3).

6.2 STAGE-STRUCTURED CONSUMER-RESOURCE DYNAMICS

Like the discrete dynamical models above, stage-structured or age-structured C-R models tend to also be accompanied by more complex dynamical phenomena than unstructured C-R models. Murdoch et al. (2003) thoroughly reviewed consumer-resource theory and argued that stage-structured and unstructured consumer-resource models have qualitatively similar results, with the difference that stage-structured oscillations are not always of the C-R variety. More specifically, stage-structured C-R models can often produce *generation cycles*. This population level cyclic phenomenon, for example, materializes when a cohort of individuals moves through the age group or stage and suppresses additional cohorts via some form of density dependence. This suppression of new cohorts remains until the large cohort dies off. With the removal of the strong cohort a new cohort is suddenly freed to thrive, which in turn suppresses another generation of new recruits.

Clearly any mechanism that prevents the creation of such powerful cohorts ought to stabilize these population level oscillations. This suggests that despite the potential for more complicated dynamics, any biological mechanism that mutes energy flux and prevents a population from increasing to overcompensatory densities ought to stabilize these excitable systems even if they are stage-structured. Thus, the coupling/interaction strength conjecture from chapter 5 may still hold.

To explore this idea let us start by considering the following simple stage-structured C-R model:

$$
\begin{cases}
\dfrac{dR}{dt} = rR(t - \tau)\left(1 - \dfrac{R(t - \tau)}{K}\right) - \dfrac{a_{\max}RC}{R + R_o}, \\[3mm]
\dfrac{dC}{dt} = \dfrac{ea_{\max}RC}{R + R_o} - mC,
\end{cases}
\tag{6.3}
$$

where all parameters have been previously defined but τ, which is the delay in time before an individual in the resource population matures into an adult. Thus, R effectively follows the adult resource population dynamics, and recruitment into the adult class comes from juveniles born τ years ago.

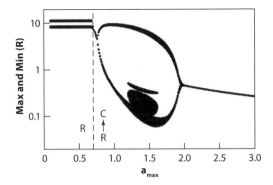

FIGURE 6.4. Local maxima and minima on the attractor of the stage-structured model are plotted as a function of an increasing maximum attack rate, a_{max}. The region categorized by R is just the R dynamics before the C has high enough attack rates to enter the population. The C-R region is also identified. The system is initially destabilized by an increased attack rate as C first enters; however, the dynamics eventually become stabilized. This stabilization is a result of the stage structure. In effect, the juvenile resource cannot be consumed by the consumer and so is decoupled from the interaction. At high enough consumption rates the resource dynamics are effectively driven solely by recruitment (which is implicitly stable in such a model). See the text for a discussion. Parameters: $r = 0.60$; $K = 10.0$; $e = 0.80$; $m = 0.50$; $R_o = 1.0$; $\tau = 2.70$.

Figure 6.4 shows the response of the adult resource population dynamics, R, to increases in attack rate where the underlying R subsystem is already displaying population cycles. This is apparent from the fact that for values of attack rate $a_{max} < 0.70$, the consumer, C, cannot survive and the R dynamics are oscillatory during this interval in a_{max}. In figure 6.4 this is evident from the two solid lines of data for $a_{max} < 0.70$. These lines represent the high and low parts of the fluctuating R population dynamics. Upon entry of the consumer, C, into the population, the dynamics are first briefly stabilized (a single curve from approximately $a_{max} = 0.70$ to $a_{max} = 0.800$) before the consumer, given high enough attack rates, eventually incurs a consumer-resource cycle, $a_{max} = 0.80$ to $a_{max} = 2.0$.

This cycle, though, does not last, and at high enough attack rates the dynamics once again are stabilized. This final result seems odd, but upon considering the biology behind the stage-structured model, the reason for this stabilization becomes clear. System (6.2) embodies the dynamics of the adult resource, which is fed on by the consumer. The adult resource is thus coupled directly to the consumer. The juvenile resource stage, on the other hand, is fully decoupled from the consumer.

At high enough consumer densities the population of adult resources is suppressed. However, the few resource adults that make it to maturity before being consumed reproduce and, in turn, are fueled by the invulnerable juvenile stage class. The relative contribution of the invulnerable, or decoupled, juvenile stage class to the resource population dynamics increases when consumer density is high—a stable influx of juveniles balances the consumption of adult resources by the consumer. This result is similar to the one we found for the type III functional response in chapter 5. In that particular case, resource refugia stabilized the C-R interaction by weakening the potentially high attack rate. Here, juvenile recruitment—decoupled from predation—acts as refugia in time (possibly in space as well if the juveniles exist in another habitat) and ultimately weakens the realized overall amount of predation on the resource.

While I have considered only resource stage structure, a similar answer is found if the consumer population is stage-structured. It is often the case that population lags are more pronounced in larger-bodied organisms. Thus, the dynamics of a C-R interaction may be more influenced in many cases by lags incurred in larger-bodied consumers. The results above readily occur again in such cases, but here it is the juvenile consumer population that is fully decoupled from the focal C-R interaction (McCann and Yodzis, 1998).

Obviously, the juvenile stages of organisms may not be expected to reproduce in such a stable fashion, even given a constant adult size. In fact, data on the recruitment of most organisms find the opposite—an enormous amount of variation in recruitment dynamics (Myers, 2001). In this case, the dynamics may be expected to follow a pattern similar to that in figure 6.4, but at high attack rates the population level variability due to recruitment variability will suddenly be fully expressed.

Another aspect of stage structure is that it is occasionally accompanied by predation between cohorts, more commonly known as *cannibalism*. Cannibalism is known to have strong dynamical implications. At the level of the population, strong cannibalism can drive density-dependent fluctuations because strong density dependence and lags are a recipe for single-population cycles (Hastings and Costantino, 1987; Costantino et al., 1997). However, cannibalism that targets newborn individuals can also stabilize potentially cyclic consumer-resource dynamics (Claessen et al., 2000). Again, this is related to the general ideas of chapter 5 in that cannibalism increases the dissipative loss terms relative to the strength of the C-R coupling terms. We expect it to be a potent stabilizing force under these conditions, and it is (Claessen et al., 2004).

Finally, stage-structured interactions allow for the possibility that changing population level traits likely feeds back to change the strength of consumer-resource interactions. As an example from lake trout ecosystems,

Zanden and Rasmussen (1999) and Vander Zanden and Rasmussen (2001) have found that lake trout, a top predator in Canadian Shield lakes, are large-bodied and piscivorous in large lakes and smaller and much more benthivorous in small lakes. The life history traits of these different lake trout populations are also different (Shuter et al., 1998). Thus, in small lakes the interaction strength for lake trout on pelagic forage fish can be expected to be much lower than that in large lakes. The modified life history attributes (e.g., smaller, more rapidly maturing lake trout in small lakes) likely alter the interaction strength of lake trout on the preferred pelagic prey, cisco.

This admits the possibility that the changing nature of life history attributes can excite or mute critical food web interactions. In the case of small lakes, there is evidence that lake trout can suppress pelagic forage fish (e.g., cisco) and become more omnivorous as a behavioral response (McCann et al., 2005; Dolson et al., 2009). This behavioral response may ultimately lead to changing life history attributes and reduced predatory pressure on cisco, ultimately mediating the persistence of the pelagic prey item, cisco. It seems possible that such feedbacks play a major role in the persistence of complex ecosystems, and so I now review literature that is beginning to show that stage structure and life history feedbacks often drive a curious C-R dynamic outcome, alternative states.

6.3 STAGE-STRUCTURED INTERACTIONS AND ALTERNATIVE STATES

Up to this point lagged or structured C-R interactions could be interpreted within the biological framework outlined in previous chapters. Stage-structured C-R interactions, though, frequently admit alternative stable states (DeRoos and Persson, 2002; Persson et al., 2007; Schreiber and Rudolf, 2008). This growing stage-structured theory has found that system level feedbacks may activate change in population structure and thereby modify food web interactions and interaction strengths. These feedbacks occasionally generate alternative steady states.

As an example, Leeuwen et al. (2008) have argued that cod-sprat interactions in the Baltic Sea create alternative states. Cod selectively predate juvenile sprat. This selective predation thins out the sprat population structure, allowing high somatic growth rates (reduced intracohort density dependence) and a population structure skewed toward large adult sprat with high reproductive rates (figure 6.5a). The highly productive sprat produce many juveniles that feed a healthy cod population. In a simplified sense, cod cull juvenile sprat produced by the relatively unconsumed stage structures (e.g., large adult sprat).

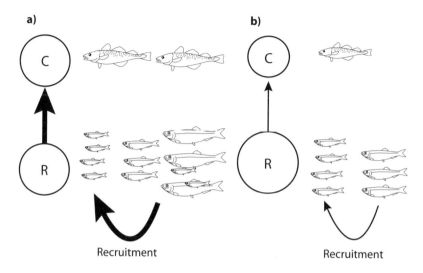

FIGURE 6.5. Schematic of cod-sprat interaction in the Baltic Sea modified. (a) Sprat populations are skewed toward producing a healthy large adult population that produces a lot of offspring that support cod. Cod thin these cohorts, reducing intraspecific density dependence and heightening sprat growth rates. (b) Sprat populations are now dominated by stunted adults that produce fewer juvenile sprat that may not be able to support a cod population. From Leeuwen et al. (2008).

Alternatively though, if cod are severely depleted—say through harvesting—the prey sprat population structure changes. Large juvenile sprat populations are no longer culled, and so juvenile sprat experience reduced growth rates and stunted adult growth forms. Thus, the population structure now becomes skewed toward stunted adult sprat that have much lower reproductive rates (figure 6.5b). In this new state the cod may not have enough prey to persist, and so an alternative state without cod develops.

Here, feedbacks operate to mediate interaction strengths in the food web and so drive an emergent allée effect for the cod. Food web level feedbacks therefore can generate a "high production" sprat system or a "low production" sprat system. Leeuwen et al. (2008) argue that this may also be the case for cod-capelin interaction in the northwest Atlantic off the coast of Canada; however, there is also evidence suggesting that capelin may not follow the required trends.

In another example from aquatic systems, where a lot of work on stage structure has been done, fisheries researchers Carl Walters and James Kitchell have argued that life history omnivory, or intraguild predation, can also admit alternative states (figure 6.6) (Walters and Kitchell, 2001). A classic

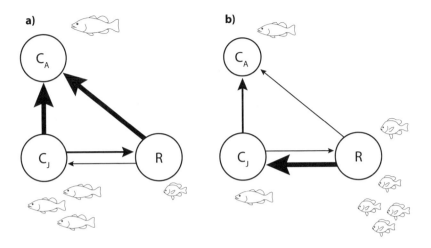

FIGURE 6.6. Schematic of a bass-sunfish interaction (Walters and Kitchell, 2001). (a) Sunfish populations are reduced by a healthy large adult population which further recruits a lot of juvenile bass because competition/predation by sunfish is diminished. (b) Once bass populations are reduced by harvesting, sunfish densities grow and increase their competitive/predatory effects on juvenile bass. This feeds back to produce weak recruitment, mediating an alternative state with sunfish dominating the littoral fish community.

example of a freshwater case involves the sunfish-bass interaction. In the presence of a healthy adult bass population, the sunfish prey are held in check (figure 6.6a). This decreased sunfish population reduces the competitive and predatory effects of sunfish on juvenile bass (figure 6.6a). Thus, in this case, the adult bass predation feeds back to facilitate high bass recruitment and so promotes a state that maintains a low-density sunfish population. However, when the adult bass biomass is depleted through harvesting, bass can no longer reduce sunfish significantly, and the negative influence of sunfish on juvenile bass increases, ultimately reducing bass recruitment (figure 6.6b). Thus, once the bass are depleted, sunfish populations can rise and hold bass populations in a low production state. The system has again reached a new alternative state with dominance by the sunfish (figure 6.6b). Importantly, even in the absence of harvesting, the bass will not return. Again, system level feedbacks mediate the strength of interactions within the food web. The food web is a dynamical entity, a lesson that will repeat itself in coming chapters.

    In summary, changing the population structure potentially greatly modifies food web interactions and/or the distributions of food web strengths. Luckily, a thorough literature on the role of stage and age structure exists for the base C-R interaction module, and some results for higher-order interactions are emerging (DeRoos and Persson, 2002; Murdoch et al., 2002; Roos et al., 2003; Persson and deRoos, 2003; Persson et al., 2007). Further, while some different dynamical outcomes emerge more readily (e.g., alternative states), here I have argued that the general results regarding local stability still remain. As an example, the role of weak or stable interactions stabilizing potentially oscillatory interactions (even if a stage-structured driven oscillation) remain, as discussed above.

6.4 EMPIRICAL RESULTS

There is little empirical work, that I am aware of, looking at how stage structure and food web interaction influence stability. In a recent example, though, Andersson et al. (2007) argued that stage-structured cannibalism acts to mute consumer-resource fluctuations in a number of empirically observed populations. There is some microcosm work by Ed McCauley and colleagues, who have illustrated how small-amplitude stage-structured consumer dynamics and large-amplitude consumer-resource cycles coexist in daphnia-algae aquatic systems (McCauley et al., 2008). Their example is an interesting one because it speaks to the stability (e.g., population variability) of these coupled subsystems and also shows an example of how structured populations can drive alternative states.

McCauley et al. (2008) use a stage-structured consumer-resource model to show that over a range of algae carrying capacity, K, the model goes through the traditional paradox of enrichment sequence (i.e., stable to large-amplitude cycles), but this same stage-structured interaction also allows an alternative small-amplitude attractor over a similar region of carrying capacity, K. Thus, the stage structure potentially influences the consumer-resource interaction in a manner that can inhibit realization of the large-amplitude cycles. That is, depending on the initial values of daphnia-algae, the dynamics can take on small-amplitude cycles instead of the large-amplitude cycle variety. In a sense, then, stage structure can act to stop the interaction from realizing the wild oscillations predicted by unstructured consumer-resource models. The experimental data of McCauley et al. (2008) amazingly reproduce these two types of cycles (figure 6.7). They further argue that the delay to maturation of the daphnia governs which type of cycle is realized. In their experiments, these small-amplitude cycles dominate 80 percent of the experiments. The more stable small-amplitude cycles rear up in this experiment when juvenile consumers have slower growth rates, which acts to reduce the amount of

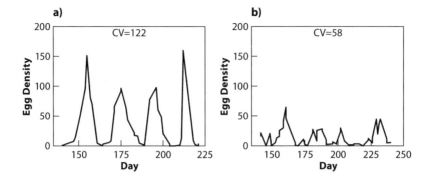

FIGURE 6.7. Egg density dynamics of daphnia displaying large-amplitude cycles (a) and small-amplitude cycles (b). Different dynamical outcomes depend on initial values. The more stable small-amplitude cycles occur when juvenile consumers have slower growth rates, which acts to reduce consumption of algae by large daphnia. The stage structure decouples or weakens this potentially potent destabilizing interaction. Modified from McCauley et al. (2008).

consumption of algae by large daphnia. The stage structure, in effect, acts to decouple or weaken this potentially potent destabilizing interaction.

Recently, Rudolf (2008) considered how stage structure modified attack rates in a predator-prey interaction. In a field experiment with damselfly and dragonfly larvae, he experimentally manipulated the density of different prey stage classes. It turned out that size structure in the prey decreased the impact of the predator on overall prey mortality by 25 percent to 48 percent at medium and high prey densities. Predation rates were effectively constant across prey densities when in the presence of a large prey class. While this does not actually show that the interaction is effectively stabilized by the changing life history, it does agree with the theory that this biology may in fact be an important structural attribute that modifies interaction strength, here reducing the predator-prey coupling strength.

6.5 SUMMARY

1. Increasing the coupling parameters (e.g., attack rates) tends to destabilize stage-structured interactions up to a point. However, if a relatively invulnerable or stable stage class exists, at some point this destabilization process is reversed. The invulnerable class or stable class begins to dominate the dynamics and acts as a stable subsidy for the other stage classes. Examples

highlight this result:

(i) if an invulnerable resource stage class exists, as the other class is suppressed by increased attack rates at some point, the invulnerable class acts to subsidize the vulnerable class, and

(ii) if a stable consumer class exists, once the resource is suppressed by the high consumption rates of the other consumer class, this stable cohort again effectively subsidizes this consumer cohort, keeping it at a steady state. Again, as the coupling strength on the vulnerable, or potentially unstable, stage class increases, eventually the invulnerable, or stable, stage classes ultimately weaken the relative strength of the increased coupling terms as their numbers begin to dominate the stage structure.

2. Stage structure promotes the possibility of alternative states whereby different stage structures are preserved by positive C-R feedbacks.

3. Stage structure can readily act to change the interaction strengths of a C-R interaction. As such, it clearly can play a powerful role in mediating food web stability or the lack of stability.

CHAPTER SEVEN

Food Chains and Omnivory

It is a natural progression for ecological theory to move beyond the consumer-resource interaction in order to explore common subsystems of food webs (Holt and Loreau, 2002). In fact, May (1974a), who championed the classic whole food web matrix approach, argued that models of intermediate complexity may be a more direct path to interpreting how food web structure influences population dynamics and stability. The notion of thoroughly exploring subsystems has evidently been around for a while. In this chapter I will begin to consider extensions of C-R theory to include simple but common three-species modules, and in the next chapter I will continue in this fashion with both three- and four-species modules. The number of possible three- or four-species modules is large, and so I will concentrate on modules that appear to be common.

The idea of studying subsystems developed from some of the seminal empirical work of Robert Paine (1980, 1992). This branch of ecology sought to understand the role of indirect effects in these complex systems. Researchers studying indirect interactions in food webs simplified the approach by focusing on "strong" interaction modules (Menge, 1995). The strongly interacting modules were experimentally determined and based on the fact that all species were significantly influenced by modifying the abundance of any of the other strongly interacting species. These strong interactions are functional interaction strengths in that they combine both direct and indirect effects (see chapter 5). In the intertidal systems studied, Menge (1995) found that apparent competition and the keystone predation module were common relative to nine other subsystem types. Here, the keystone predation module is closely related to the diamond module in that it includes a generalist predator that forages on multiple competing prey. Despite the different definition of interaction strength, the answers resonate with more recent whole web approaches.

Network analysts ask precisely the types of questions broached by recent theory for a variety of modular networks (Milo et al., 2002). Food webs have found their way into this literature, and the results are most useful. As an example, network analysts decompose whole ecosystems into the set of all

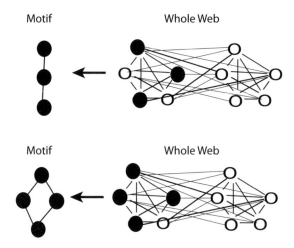

FIGURE 7.1. Food webs can be deconstructed into subsystems or motifs. Here, two common motifs are shown. Having a common structure means that these motifs appear with greater frequency than expected by pure chance. Different types of networks have different types of ubiquitous structure. It has been found, for example, that information-processing networks are different in structure from energy processing–based networks (Milo et al., 2002). Figure modified from Carlos Melian's depiction of motifs (http://press.princeton.edu/titles/xxxx.html).

possible subsystems or motifs (see figure 7.1 for examples of motifs). Network analysts often call these motifs because they do not follow interaction strength, but here I will continue to consider them modules, as we will soon investigate the role of the strength of these interactions as well. Once the whole web is decomposed, analysts assess the ubiquity of these modules by comparing them to randomnly created webs. Modules that are significantly more common than random are overrepresented and seen as major building blocks of the system. Here, I will refer to them as fundamental modules. Figure 7.1 highlights two common and, arguably, fundamental modules behind the construction of whole webs (Milo et al., 2002; Bascompte and Melian, 2005; Camacho et al., 2007; Stouffer and Amaral, 2007).

In what follows, I will look at food chain and omnivory modules, which of all three species modules in food webs appear to be the most often over-represented (Milo et al., 2002; Bascompte and Melian, 2005; Stouffer and Amaral, 2007; Camacho et al., 2007). In the next chapter I will consider four species modules, as well as spend some time on the role of generalism (effectively a three-species module with a consumer feeding on two resources). All the while, though, I will pay attention to existing C-R theory in order to interpret the consequences of these larger modules. These food web modules have a tendency to display subsystem signatures such that oscillations can be attributed to particular C-R interactions [e.g., Muratori and Rinaldi (1992)]. Thus, we will approach this problem by looking at how interaction strength mediates the dynamic behavior of the underlying consumer-resource interactions. Clearly, the ultimate goal is to understand modules within a whole system framework, allowing us to bridge modular results with classic theory (a topic of a later chapter).

From the previous chapters we have found that we can understand the dynamics of C-R interactions by understanding the fluxes through interactions. Recall that a C-R interaction with a strong coupling term (i.e., lots of flux through the C-R interaction) relative to the loss terms tends toward instability. We can conjecture that the least stable interaction within a subsystem (i.e., a potentially oscillatory component) may be stabilized if placed within a food web structure that acts to reduce the relative coupling strength of this potentially unstable interaction. Different subsystems may act to constrain a potentially unstable subsystem. Similarly, a relatively stable interaction may be excited by a food web structure that increases the focal consumer's relative coupling strength on a given resource.

Here, I begin to consider some common simple modular food web structures and ask if the dynamics of subsystems can be seen using the framework laid out in previous chapters. Specifically, I seek answers to the following questions: (1) When does common food web structure increase or weaken the relative interactions strengths (i.e., coupling terms relative to loss terms), and/or (2) When does a food web structure modify flux between consumers and resources in a density-dependent manner such that the food web tends to increase flux rates in some situations (say, when a resource attains a high density) and decrease the coupling in other situations (e.g., when a resource falls to low density). Answers to these two questions pave the way for generally understanding how food web structure mediates the relative dynamics and stability of different food web configurations. In this chapter, I explore a food chain (two coupled C-R interactions), as well as omivory. Finally, I will once again consider how stage structure can influence this food chain theory.

7.1 A FAMILIAR MODULAR EXAMPLE: FOOD CHAINS

There is an existing argument that long food chains are inherently unstable in the sense that increased trophic length tends to beget instability and collapse (Pimm, 1982; Pimm and Kitching, 1987). Additionally, food chain models readily generate complex dynamics, and an industry of research has been put into showing this (Hastings and Powell, 1991; Muratori and Rinaldi, 1992; McCann and Yodzis, 1994, 1995; Kuznetsov and Rinaldi, 1996). In an informative contribution, Muratori and Rinaldi (1992) showed that highly efficient (i.e., strong) interactions tended to produce cycles in the underlying C-R subsystems of a food chain. This, in turn, produced complex or chaotic dynamic behavior. This result hints at the fact that the arrangement of interaction strengths likely plays a major role in mediating population dynamics. To investigate this further, we start with the simple L-V extension to a three-species food chain:

$$
\begin{cases}
\dfrac{dR}{dt} = rR(1 - R/K) - a_C RC, \\[2mm]
\dfrac{dC}{dt} = ea_C RC - m_C C, \\[2mm]
\dfrac{dP}{dt} = ea_P CP - m_P P,
\end{cases}
\tag{7.1}
$$

where the parameters are as defined before with attack rates of a predator/consumer on its prey/resource, i, identified by the subscript i, and the mortality rate of the population also identified by the subscript i.

To start, let us hold the parameters constant for the consumer-resource interaction and modify only the predator-consumer attack rates, a_P. In this case we start with a C-R interaction where the dynamics are locally stable but show overshoot potential (i.e., the dominant eigenvalue is complex). As we increase the predator attack rate, we reach a point where the predator can coexist. This happens at the attack rate $a_P = 0.30$ in figure 7.2. Notice that at this point the dominant eigenvalue is 0.0. This tells us there is a bifurcation, and this coincides with the entry of the predator into the system. Recall that the dominant eigenvalue is the one with the largest real component and so determines the local stability of the equilibrium. For comparative purposes, I have also included a dashed line that shows the dominant eigenvalue for the C-R interaction in isolation. Notice that as this predator attack rate is increased, we see that for relatively weak attack rates the dominant eigenvalue for the P-C-R system (solid curve) drops below the C-R eigenvalue. This marks the region in parameter space where the predator actually stabilizes the system relative to the C-R subsystem in isolation ($a_P = 0.56$ to 3.0). However, as attack rates increase still further, the predator actually again destabilizes the system relative to the C-R interaction. The result, relatively weak to intermediate coupling strengths of the predator, stabilizes the underlying excitable interaction. Strong predatory coupling further destabilizes the interaction. This is consistent with recent theory (McCann et al., 1998).

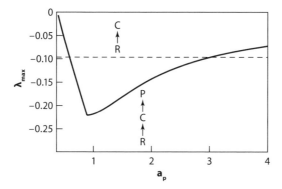

FIGURE 7.2. Local stability response of a Lotka-Volterra food chain model under an increasing attack rate, a_P, by a predator on the intermediate consumer, C. The now familiar qualitative result of stabilization followed by destabilization reappears. The underlying C-R eigenvalue (dashed line) is included for comparison. Clearly, a relatively weakly interacting predator can stabilize the underlying C-R interaction. Effectively, in this region the weak P-C interaction increases the loss term on the C-R interaction. As shown in chapter 5, this tends to dampen an excitable interaction. Parameters: $r = 1.0$; $a_C = 0.62$; $e = 1.0$; $m_C = 0.5$; $m_P = 0.20$; $K = 4.0$.

Mechanistically, this result resonates with the conjectures of C-R theory. Here, although the relative strength of the predator-consumer interaction strength is increasing, it does not act to destabilize the overall food web until it is quite strong (e.g., $a_P > 3.0$). Prior to that the interaction acts to modify the realized flux rate through the C-R subsystem. For relatively weak to intermediate strengths, it effectively acts to dissipate energy from this potentially excitable C-R interaction and so stabilizes the food chain.

To this point I have emphasized the local stability aspects of modules and pairwise interactions. I did this because McCann et al. (1998) employed, and emphasized, largely nonequilibrium metrics (e.g., the coefficient of variation or local minima and maxima of nonequilibrium steady states) to show how weak interactions stabilize food webs, and here I wanted to emphasize that the results extend to local equilibrium stability as well as variability.

The eigenvalue experiment, though, estimates the local stability of all members of the community. As such, the maximum eigenvalue sometimes can tend to reflect one particularly unstable aspect of the community. For example, in the above two cases the eigenvalue at a low a_P starts at zero. This zero eigenvalue reflects the value at which the top predator can finally invade this system (this is called a *transcritical bifurcation*). Thus, the zero eigenvalue speaks to the predator but does not indicate what the invasion of the predator does to the underlying C-R interaction. We might also want to consider how this predator invasion influences the underlying C-R interaction. One way to approach this theoretically is to examine the dynamics of the C-R interaction (not the eigenvalue) across all values in the predator attack rate.

As an example we will now consider the R-M three-species version of the food chain. This model produces rich dynamical behavior and is more informative in this sense than the classic L-V model. The Rosenzweig-MacArthur version of the food chain can be written as

$$
\begin{cases}
\dfrac{dR}{dt} = rR\left(1 - \dfrac{R}{K}\right) - \dfrac{a_{C,\max}RC}{R+R_o}, \\[2ex]
\dfrac{dC}{dt} = \dfrac{ea_{C,\max}RC}{R+R_o} - m_C C - \dfrac{a_{P,\max}CP}{C+C_o}, \\[2ex]
\dfrac{dP}{dt} = \dfrac{ea_{P,\max}CP}{C+C_o} - m_P P,
\end{cases} \qquad (7.2)
$$

where all the parameters are as above but with the additional half-saturation parameters of the type II functional response, R_o, and C_o.

To explore the dynamics we plot the local maxima and minima of the attractor of the consumer's densities for any given value of $a_{P,\max}$. Again, we start with the scenario where the underlying C-R interaction is already excited (this time actually oscillatory) and then increase the predator attack rate.

As in figure 7.2, the top predator cannot gain entry into the community until its attack rate is sufficient to balance its loss rate. At approximately $a_{P,\max} = 1.0$, the predator enters the community, and immediately we find that the underlying C-R oscillation (indicated by the two lines of points up to $a_{P,\max} = 1.0$) becomes a single line of points. This indicates the presence of a stable equilibrium once the predator invades. Thus, weak predator attack rates stabilize the C-R interaction relative to the case without the predator. Again, the weak attack rates of the predator act to increase the dissipative loss of the consumer and so weaken its relative coupling strength on the resource. However, this is short-lived in this example, and with increased attack rates the equilibrium soon becomes unstable and shows much more variance than the underlying C-R oscillation alone (for $a_{P,\max} > 1.40$). The data in figure 7.3 show some period-doubling activity indicating that the dynamics become more complex than a simple cycle as the predator attack rate is increased. Thus, the coupling strength relative to the dissipative loss of the P-C interaction eventually increases enough such that it is fully destabilized. The P-C interaction effectively adds another oscillator to the food chain, which then interacts with the C-R oscillator, generating more complex dynamical phenomena. In fact, in this model it is easy to get chaotic dynamics. Hastings and Powell (1991)

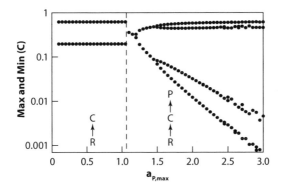

FIGURE 7.3. Bifurcation diagram of the R-M food chain model under increasing $a_{P,\max}$ by a predator on an intermediate consumer. Population variability of the consumer is followed by keeping track of local maxima and minima. Again, even with this slightly different stability metric, the now familiar qualitative result of stabilization followed by destabilization reappears. Parameters: $r = 1.0$; $K = 2.0$; $R_o = 0.50$; $e = 0.50$; $m_C = 0.50$; $C_o = 0.30$; $m_P = 0.30$; $a_{C,\max} = 1.80$.

showed that chaos readily emerges when the attack rates are high and the mortality rates are low—in other words, when both relative coupling strengths are strong in the P-C and C-R interactions.

Now, let us turn to an alternative starting condition to show that this result is general. Specifically, let us start with a C-R interaction that does not oscillate in isolation (i.e., a relatively moderate attack rate, $a_{C,\max} = 1.50$). We then choose a P-C interaction ($a_{P,\max} = 1.80$) that is strong enough to drive a cycle (in a sense, then, this cycle is due to the P-C interaction). Figure 7.4 shows this example. The conceptual experiment here is as follows: Given a P-C oscillation, we ask what happens if we weaken the C-R interaction by moving to the left on the x-axis of figure 7.4. This figure shows that as the C-R interaction is weakened, the potentially oscillatory P-C interaction is stabilized. In a sense, weakening the C-R interaction cascades up the food chain to weaken the P-C interaction, as the weakened C-R effectively means less flux ultimately also goes through the P-C interaction. Therefore, both coupling terms are weakened by this action, and so the relative strengths of the C-R and P-C interactions are diminished, driving stable equilibrium dynamics. Again, the result resonates with previous chapters.

Above, I have shown the results from several conceptual experiments. The results are in fact robust, and this can be seen by examining the underlying geometry of the isoclines for the food chain model above. McCann and Yodzis (1995), for example, performed a bifurcation analysis of system (7.2) using

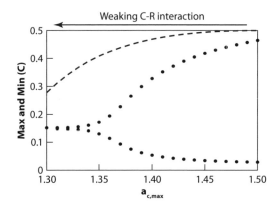

FIGURE 7.4. Local stability response of a Rosenzweig-MacArthur food chain model across a range of the intermediate consumer attack rate, $a_{C,\max}$, on the resource. We start with a strong predator attack rate ($a_{P,\max} = 1.80$) that produces P-C cycles for C's maximum attack rate ($a_{C,\max} = 1.50$). If we move from right to left, in other words tune down the consumer attack rate, we effectively remove the P-C oscillation. In a sense, reducing $a_{C,\max}$ cascades up to reduce the flux rate through the P-C interaction and so stabilizes the whole chain. Parameters: $r = 1.0$; $K = 2.0$; $R_o = 0.50$; $e = 0.50$; $m_C = 0.50$; $C_o = 0.30$; $m_P = 0.30$; $a_{C,\max} = 1.50$.

the isocline geometry as a guide for understanding the qualitative changes in dynamics. This is beyond the scope of this book, but the results show that the above conceptual experiments are independent of parameter choice. They also show that multiple basins can arise in simple food chains. Here, though, the multiple basins tended to require quite strong C-R and P-C interactions. Instability at one level (a high variation in population dynamics) appeared to coincide with the onset of another form of instability, multiple basins of attraction (one of which does not include the top predator, P).

7.2 OMNIVORY

Having explored the dynamics of food chains, we now proceed to consider the case where the top predator, P, can also consume the resource. There is a lot of work on the role of omnivory in food webs, and the arguments for its role at first appear equivocal (Pimm and Lawton, 1978; Fagan, 1997; McCann and Hastings, 1997; Vandermeer, 2006). Here, I will argue that omnivory can indeed be quite stabilizing, but it is contingent on the relative strengths of the interactions. All recent and historical results can be understood from this perspective, although much confusion exists.

As an example of the changing influence of omnivory on stability, let us examine the following omnivory model:

$$
\begin{cases}
\dfrac{dR}{dt} = rR(1 - R/K) - a_C RC - a_{PR}PR, \\[2mm]
\dfrac{dC}{dt} = ea_C RC - m_C C - a_{PC}CP, \\[2mm]
\dfrac{dP}{dt} = ea_{PC}CP + ea_{PR}PR - m_P P,
\end{cases}
\tag{7.3}
$$

where the parameters are as defined before with the attack rates of the predator, j, on its prey/resource, i, identified by the subscript ji and the mortality rate of the population, k, is identified by the subscript k.

I have now switched back to the Lotka-Volterra model, but the results that follow can just as easily be obtained using other forms of food web models [e.g., McCann and Hastings (1997)]. I have chosen to return to the Lotka-Volterra form to highlight how omnivory, for different interaction strengths, can act to remove the excitability in a C-R interaction (i.e., remove the complex portion of the dominant eigenvalue and so remove overshoot population dynamics), as well as ultimately add back the excitability in the food web as the omnivorous interaction increases in strength. This result has been documented for nonequilibrium behavior using bifurcation diagrams (i.e., local minima and maxima of nonequilibrium steady states); however, it can be shown that the eigenvalue responses for the equilibrium results often mirror those for the nonequilibrium results.

Figure 7.5a shows a case where the omnivory strength of the top predator in system (7.3) is increased from a zero coupling strength on the resource (i.e., a food chain) to a strong coupling strength on the resource. Again, I have done this by increasing the top predator's attack rate on the resource (i.e., increased a_{PR}). Included in this figure is a dashed line that represents the size and existence of complex parts in any of the eigenvalues. Here, the parameter values have been chosen for the food chain (i.e., when $a_{PR} = 0$), such that the dynamics of the nonomnivorous chain can just maintain the predator.

As we increase the omnivory strength, we see that omnivory is stabilizing in that it drives the eigenvalue to more negative values (solid curve). Increasing omnivory in this case allows a barely feasible population of top predators, P, to increase its density away from near-zero densities. Mathematically, the increased consumption of resources pushes the predator further away from the transcritical bifurcation where the dominant eigenvalue is zero and real. Additional flux to the predator does not drive overcompensatory dynamics but rather increases the predator's rate of return without the negative consequence of overshoot dynamics, as discussed in earlier chapters. This is a

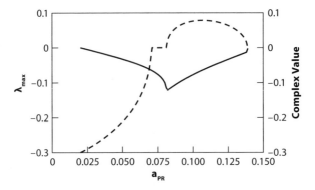

FIGURE 7.5. Local stability response (solid curve) of a Lotka-Volterra food chain model as we increase the amount of omnivory. A dashed curve shows the value of the complex part of the dominant eigenvalue. Note that the complex portion is reduced during the stabilization phase, suggesting overcompensatory stabilization due to omnivory. It disappears entirely before reappearing with the onset of destabilization, as expected. Parameters: $r = 1.0$; $K = 2.0$; $a_C = 0.20$; $a_{PC} = 0.30$; $e = 0.80$; $m_C = 0.50$; $m_P = 0.20$.

form of stabilization from omnivory that I have not seen discussed in food web ecology. Here, I will refer to this as *feasability stabilization* because the added omnivorous contribution allows the predator to maintain densities bounded away from dangerously small densities. Note that this form of stabilization ought to occur when the dominant eigenvalue is real and near zero. Increased flux to the top predator then stabilizes the food chain, as discussed in earlier chapters.

Omnivory can similarly also stabilize overcompensatory interactions. In the case above, even when the dominant eigenvalue is real, the two other eigenvalues are complex (likely due in a simplified sense to the C-R interaction, which is relatively strongly coupled). Omnivory from the top predator can act to shunt some of the energy away from this potentially excitable C-R interaction (i.e., energy is deflected away from the C-R interaction and up the P-R pathway) and, in doing so, can weaken the relative coupling strength of the C-R interaction. I will refer to this as *overcompensation stabilization*. Figure 7.5 shows this complex part (dashed curve) of the subdominant eigenvalue, and one can see that it is initially reduced by increased omnivory. In this case, relatively weak to intermediate omnivory acts to increase the dissipative loss on the resource and so reduces the relative interaction strength of the underlying C-R interaction. Clearly, as the P-R interaction is further increased, omnivory becomes destabilizing in a dramatic sense as the predator eventually outcompetes its prey item (C) and so eliminates C.

One can conceive of a third case that lies outside the above results. We can create a scenario where omnivory is fully destabilizing to a food web regardless of its strength. If we choose parameters such that the intermediate consumer, C, starts at dangerously low densities (i.e., close to a transcritical bifurcation where C becomes 0), increased omnivory, at no cost, elevates the top predator density and thus pushes the intermediate consumer out of existence. There is no stabilization phase in this case. Again, this occurs when the dominant eigenvalue is real and relatively close to zero (i.e., very close to the transcritical bifurcation where C goes from positive to zero densities with increases in omnivory). I will call this *omnivory suppression*. Curiously, if the subdominant eigenvalue (i.e., not the largest eigenvalue) is complex in such a situation, omnivory may reduce the size of this complex part, as it usually does; however, because it is not dominant, this stabilizing effect is outweighed by the removal of an already weak intermediate consumer, C. If I were to instead create such a scenario as above (i.e., C low) in a population oscillating and followed maxima and minima, I would find that omnivory tends to inhibit the amplitude of the oscillation until suddenly C is suppressed out of existence (this is known as a cyclic transcritical bifurcation). In this case, the maxima and minima hide the fact that C is moving precariously toward zero density. Nonetheless, both results resonate with each other.

It is easy to reproduce the above results by choosing parameters that beget the three cases: (1) a predator population near zero in the food chain (feasability stabilization); (2) a food chain that has a complex dominant eigenvalue (overcompensation stabilization) with a C not critically close to zero, and (3) a food chain with critically low C and positive P values (omnivory suppression). In the first two cases increasing omnivory first stabilizes and then destabilizes. The destabilization phase can be due either to the fact that the predator—once capable of high consumption rates—can drive C to extinction (i.e., push the dominant eigenvalue through a transcritical bifurcation) or this increased capacity of the predator can destabilize the food web in the sense that it can contribute a new excited interaction that produces overshoot dynamics or oscillations [e.g., (McCann and Hastings, 1997; Tanabe and Toshiyuki, 2005)]. The final case, as mentioned, will produce only destabilization (Vandermeer, 2006).

Above, I have attempted a synthesis of the role of omnivory in stabilization. However, the theoretical literature on this topic appears murky and unclear. Some reconcilatory comments are in order. In a recent paper, Tanabe and Toshiyuki (2005) showed that omnivory in the Lotka-Volterra model can lead to chaos. This was interpreted by the authors as a different result from that found by McCann and Hastings (1997). While interesting,

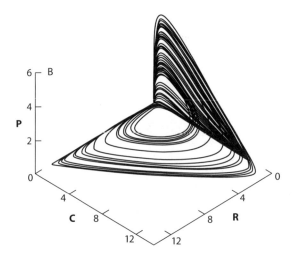

FIGURE 7.6. Dynamics of the Tanabe and Toshiyuki (2005) omnivory model (same as above model) on the phase space. Note that the dynamics trace out the two underlying oscillators, one parallel to the C-R plane and one parallel to the P-R plane. Modified from Tanabe and Toshiyuki (2005).

especially because it occurs for type I feeding relationships, this result is not unexpected because increasing omnivory strength should be expected to lead to the addition of another excited interaction (i.e., the strong P-R interaction eventually becomes an oscillator) that can inspire chaos when coupled to other excited interactions (e.g., the C-R interaction). Figure 7.6, from Tanabe and Toshiyuki (2005), shows exactly this—the trajectory of the chaotic attractor carves out a pathway that highlights two clear underlying oscillators. The strange attractor is clearly composed of two parts: One part is the oscillator traced out parallel to the P-R plane, and the second part is another oscillator that traces out a cycle parallel to the C-R plane (figure 7.6). This result highlights the fact that a strong omnivorous interaction can produce complicated dynamics when coupled to a strong C-R interaction (McCann and Hastings, 1997).

Similarly, Vandermeer (2006) argued that omnivory can both stabilize and destabilize dynamics. He further, and importantly, noticed that omnivory sometimes can be entirely destabilizing regardless of strength. This result is true and is consistent with the arguments above, but it remains important to emphasize that the resulting stabilization, or lack of it, is far from random. Weak omnivory tends to stabilize, while relatively strong omnivory destabilizes. In cases where the consumer is close to zero (the omnivory suppression case), omnivory does not stabilize the dominant eigenvalue but rather rapidly drives the consumer out of existence. As mentioned above, though, a cursory exploration suggests that any underlying complex portion of the eigenvalues in the latter case tends to be reduced as one increases omnivory. Omnivory thus has two major tendencies: (1) when relatively weak, it can mute over-compensatory food chain potentials. In such cases it acts as a dissipative force and weakens the relative strength of potentially strong and oscillatory subsystems. (2) Omnivory also has the potential to strongly suppress its prey and competitor, C, and so it also has the potential to be a great destabilizing factor (Vandermeer, 2006).

I now turn to the classic results of Pimm and Lawton (1978). These results are important to consider because these investigators employed different techniques to address this problem and rigorously showed that in the set of all omnivory strengths there is only a relatively small region that gives a stable equilibrium. As such, on odds alone, one would expect little omnivory in the world. There are two ways to interpret this result. One is that the answer Pimm and Lawton (1978) got is different because their method was a randomization method that employs general matrices, which just give different results. The other is that the results are not in conflict if we examine them a little more carefully. I would argue that the latter is correct.

A student of mine, Gabriel Gellner, has done an analysis that integrates the classic matrix approach with the bifurcation style approach of modular theory (Gellner and McCann, forthcoming). He has derived a fairly simple methodology that uses the classic matrix approach but modifies its analysis to be consistent with the modular theory bifurcation approach. In other words, the researcher varies the strength of omnivory and asks whether omnivory is stabilizing or destabilizing relative to the food chain case (i.e., no omnivory). When he does this, he generally finds the familiar result of stabilization followed by destabilization discussed above—thus, relatively weak interactions appear to be generally stabilizing even using the results from Pimm and Lawton (1978). Recall that they found that for those omnivorous models that were stable, return times tended to be shorter and so more stable than in the food chain model. The stabilization followed by the destabilization phase of modular theory produces this result.

But how does one account for the results of Pimm and Lawton (1978) regarding the probability of finding a stable web with omnivory? Gellner and McCann (forthcoming) created two metrics to assess this. A classic metric that assessed the probability of finding an omnivorous web that is more stable than the underlying food chain (probability of a more stable web, PMSW) within a given parameter range, and a modular metric that asked how the relative strength of omnivory influenced stability. Gellner and McCann (forthcoming) then reconstructed Pimm and Lawton's experiment by choosing many random webs from their biologically constrained parameter space. The results agree both with Pimm and Lawton (1978) and McCann and Hastings (1997). Under the parameter constraints of Pimm and Lawton (1978) omnivory is most often destabilizing (the destabilization phase is large); however, for weak omnivory the food web is generally almost always stabilized. The brief stabilization phase generates a small region of parameters, where the omnivorous interactions are weak, that always produces more stable webs than the isolated food chain, as Pimm and Lawton (1978) found.

If omnivory is such that an organism must reduce the intake of its preferred prey (here, C) to consume the resource, R, then omnivory remains stabilizing (McCann and Hastings, 1997; Vandermeer, 2006). If it does this in a density-dependent manner, this form of omnivory can be a potent source of stabilization (McCann et al., 2005). Strong suppression in the focal food chain creates a scenario where C is small and R is at high density [i.e., a classic trophic cascade; for a review of trophic cascades, see Carpenter and Kitchell (1996); Schmitz et al. (2000)]. Thus, the behavioral and evolutionary pressure for omnivory when interactions in the focal chain are strong becomes quite real. Once the top predator turns its attention toward R and away from C, then C is released from its suppression by P and can then increase. This creates the ability for the the top predator to drive C and R out of phase, a well-known form of stability for any generalist consumer.

This behavioral adaptability prevents sustained strong interactions on a given resource and so is powerful in muting underlying oscillators in the food web. The key to such stabilization is that an interaction strengthens at a time when a given resource is relatively high in density (culled, preventing overshoot) and weakens at a time when the resource is relatively low in density, freeing the resource to grow when it must in order to avoid extinction. Considering the omnivorous situation described, this is exactly what dynamically unfolds for a reasonably behaving top predator feeding on both C and R. Here, by reasonably I mean that the top predator switches at times that tend to enhance its growth rates. We will return to this idea of behavioral adaptability in more detail in later chapters.

7.3   STAGE STRUCTURE AND FOOD CHAIN STABILITY

Until this point I have argued that, at least qualitatively speaking, stage structure does not change our results from C-R theory. That is, a weak to moderate interaction can mute a generation cycle in a stage-structured resource

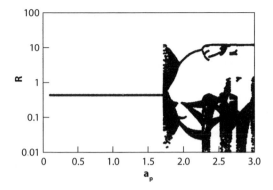

FIGURE 7.7. Bifurcation diagram of resource densities as a function of predator attack rate for a P-C-R model with a stage-structured resource [see chapter 6, model (6.3) but extended to include an unstructured predator, P]. Increased predator attacks eventually suppress consumers and release R to flourish ($a_P = 1.70$), at which point the resource produces generation cycles or cycles of the single-population variety. Parameters: $r = 0.60$; $K = 10.0$; $\tau = 4.0$; $e = 0.80$; $m_C = 0.50$; $a_C = 1.50$; $R_o = 1.0$; $C_o = 1.0$; $m_P = 0.150$.

population. Here though, as we move to higher-order modules, we need to be a little more careful as the results can be different.

As an example, let us consider a stage-structured resource fed on by a consumer [model (6.3) in chapter 6]. For a consumptive rate $a_C = 2.0$, the C-R populations alone are stable. Recall from chapter 6 that this is an example of a consumer damping out the underlying resource generation cycle. Now if I run a numerical experiment that adds a top predator, an interesting but understandable result emerges. For weak to moderate P coupling strengths (i.e., the P-C interaction does not have an underlying oscillating component), the consumer should be suppressed relative to the C-R model alone. Further, this reduction in C should cascade down such that it releases R, allowing it to attain higher R densities. With the attainment of higher R densities, one expects that the resource will eventually regain generation cycles (i.e., population growth rates and density increase and this excites generation cycles).

As an example, I changed the coupling strength of the top predator by changing its attack rate, a_P (figure 7.7). As soon as the predator can invade the system (i.e., $P > 0$), the resource immediately begins to oscillate as predicted. Further increases in predator attack rate ultimately add another oscillator to the mix, and the dynamics soon enter complex regimes (figure 7.7, $a_P > 2.50$).

Although not shown here, a stage-structured consumer population, C, experiencing a generation cycle, instead of the resource as above, can be stabilized by the above predator. The above results make sense but suggest that stage structure adds another potentially oscillating component that one must consider. A number of researchers have made some excellent contributions to the theory of stage-structured dynamics, and more work on the stabilizing or destabilizing influence remains, but these results suggest that they appear to operate consistently within the framework outlined in previous chapters.

Here, I have shown that a predator's influence cascades through the consumer to release the resource. The resource, freed to reach high densities, then undergoes stage-structured oscillations. Again though, as discussed before, the predator may be inspired to feed omnivorously in the face of such high densities of resource. Omnivory—at least omnivory that is not too strong—can again cull the resources, limiting their high densities, and so modest amounts of omnivory could stabilize the stage-structured fluctuations in this context as well.

7.4 EMPIRICAL RESULTS

While there is much research on trophic control in food chains, a sparse amount of work has been done on the dynamics of chains with and without predators. This is likely because theory has essentially argued that food chain length tends to be destabilizing (Pimm et al., 1991). While this can be true, it is also possible that predation mediates food chain persistence.

As an example, Otto et al. (2007) used empirical body size relationships from Yodzis and Innes (1992) to parameterize five empirical food webs. They created contours in the parameter space of food chain dynamics that delineated three different regions of dynamics: (1) highly oscillatory dynamics, (2) exclusion, and (3) stable persistence. These contour maps had only about 20 percent of this biologically plausible parameter space containing systems that were stable and persistent. Rather amazingly, all data from the five webs tended to fall within, or at least close to, this restricted persistence region. The size of this persistence region is clearly dependent on the magnitude of some of the underlying and largely unknown parameters (e.g., the half-saturation rate and the carrying capacity), but Otto et al. (2007) argue that their general result is robust to variations in these key parameters. This suggests that body size relationships and attack rates may constrain food webs to reasonably well-behaved dynamics.

It is worth considering historical findings in light of the above modern synthesis. Pimm and Lawton (1978) did a preliminary empirical analysis of the amount of omnivory in real systems. They compared their findings relative to the amount of omnivory one would expect by chance alone and found that the amount of omnivory in real systems is in fact much less than that expected by chance. However, Pimm and Lawton (1978) excluded weak interactions in their analysis (all omnivorous interactions <20 percent). Their results are therefore exactly what the above theory would predict—that relatively strong omnivorous interactions should rarely exist in nature.

This above theoretical synthesis, though, would also predict that omnivory would be common once we included all links. More recent analyses that include all interactions seem to agree with this (Dunne et al., 2004). Polis (1991), for example, found that 76 percent of the species in the Coachella Valley food web showed evidence of omnivory, while Sprules and Bowerman (1988) showed its prevalence in pelagic aquatic ecosystems. In a largely anecdotal contribution, Polis et al. (1996) argued cogently for the ubiquity of omnivory, both within the focal chain and across food chains (we will discuss this aspect of food webs later). Network researchers have similarly found that the food chain and the classic three-species omnivore modules are frequently overrepresented, but there are some cases where omnivory is also underrepresented [e.g., Bascompte and Melian (2005)].

In a recent meta-analysis, Thompson et al. (2007) analyzed 58 food web sets to see where omnivory resided in food webs. Their result indicates that, with the exception of plants and herbivores, omnivory is everywhere. As one moves up the food web, the degree of omnivory increases dramatically at the predatory trophic levels. Omnivory appears most common in marine systems, is rarest in streams, and is intermediate in lakes and terrestrial food webs. But little has been done, that I am aware of, to follow up on the historical empirical result of Pimm and Lawton (1978) that, while omnivory may be rampant, moderate to strong omnivory is rare. As it stands, it appears as though this is a distinct possibility. Much more work needs to be done on this, though, including considering the role natural variability plays in the amount and strength of omnivory.

It remains to address empirically what the dynamical implications of omnivory are for food webs. In recent years, some ecologists have begun to address this experimentally. In a sophisticated microcosm experiment, Holyoak and Sachdev (1998) used protists to develop a modest gradient in omnivory and then ran experiments over this omnivory gradient while keeping track of stability (e.g., persistence time, coefficient of variability).

Holyoak and Sachdev (1998) found that omnivory did seem both to enhance persistence (the length of time all species lasted in the experiment) and to reduce the variability (coefficient of variation). Figure 7.8 shows the persistence time across a gradient in omnivory for the predators. While in this experiment prey bacteriovores showed no real pattern with increased omnivory

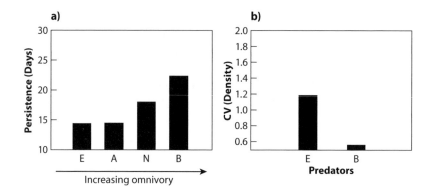

FIGURE 7.8. (a) Persistence times of predators in three trophic level systems. (b) Temporal variability (coefficient of variation, CV) of species with different degrees of omnivory. Moving from left to right tends to increase omnivory, and with it, persistence times increase and temporal variability decreases. See Holyoak and Sachdev (1998) for details. Figure modifed from Holyoak and Sachdev (1998).

(Holyoak and Sachdev, 1998), the predators showed increased persistence times (figure 7.8a) and reduced temporal variation in population density (figure 7.8b).

In another experiment, Fagan (1997) used an arthropod assemblage at Mount Saint Helens and manipulated the amount of omnivory by changing the proportion of generalists in the web. He then subjected a specific component of the assemblage to a major perturbation. In a sense, he was performing the empirical version of the thought experiment employed by MacArthur (1955), who argued that generalists garnered stability by spreading out their risk over multiple pathways. Thus, if one pathway is altered severely, a generalist species is buffered by being able to tap into alternative pathways. Fagan's (1997) experimental perturbation (an aphicide) directly altered a specific set of pathways, while not perturbing alternative pathways. These analyses found that the aphicide application increased among-species variability in per capita rates of change in plots dominated by specialists but did not significantly affect this measure in plots dominated by generalist omnivores.

This result agrees that omnivory may be stabilizing but also points out that some of the greatest relevance of omnivory, indeed generalism, is the ability to deal with variability. This aspect of omnivory needs further empirical and theoretical work. I wonder, for example, whether the amount and strength of omnivory change with the amount of environmental variability.

In a contribution that broke down a whole Caribbean marine food web into modular components, Bascompte and Melian (2005) found that the co-occurrence of strong interactions on two consecutive levels of food chains took place less frequently than expected by chance. This result suggests that destabilization of food webs by strong linear chains was not frequently found in this web. Further, wherever such a combination of strong food chain interaction strengths was found, it tended to also be accompanied by an omnivorous link, although this was found to be relatively strong as well. Similarly, Neutel et al. (2002) and Neutel et al. (2007) have examined a number of soil food webs, using both empirical data on feeding interactions and theoretical analysis, and found that weak omnivorous loops play a fundamental role in stabilizing these complex below-ground networks. The weak omnivorous interaction may be acting to dampen the strong linear food chains, as modular theory suggests. I will return to these fascinating results in more detail later, when I begin to consider whole food webs.

7.5 SUMMARY

1. In a P-C-R food chain, a weak predator-consumer interaction, when coupled to a strong consumer-resource interaction, tends to be stabilizing relative to the strong isolated C-R reaction alone. This result can be shown for return times (or eigenvalues) and in terms of population variability. The weak interaction acts as a dissipative loss to the C-R interaction, thus weakening its relative interaction strength.

2. A recent empirical analysis by Otto et al. (2007) showed that body size distributions tended to generate allometrically determined parameters that generate persistent food chains.

3. Relatively weak omnivory tends to be a stabilizing force, while strong omnivory tends to be quite often destabilizing—so much so that a fairly large region of space for certain parameter constraints does not allow coexistence.

4. The point at which an omnivorous interaction becomes destabilizing depends critically on the parameters. Under Pimm and Lawton's constraints, it does not take much omnivory to drive the destabilization phase. So, omnivory—if these constraints are real—would be restricted to quite weak scenarios. If these constraints are not so strong, it is possible that the stabilization region of omnivory may be more dramatic. This aspect of omnivory requires further empirical treatment.

5. An example of how stage structure can play into this theory is given. Specifically, it is shown that a stable C-R interaction is destabilized or excited by

the introduction of a predator. This predator reduces the consumer and in doing so frees the resource to attain high densities. Stage structure in the resource drives a cycle. Thus, the cascading influence of the predator on the consumer frees the resource to experience strong overcompensation and oscillatory dynamics. This novel result is shown to argue that ultimately food web theory must expand to consider how interactions, and populations in and of themselves, are excited or muted by the food web structure. An example of a consumer stage structured cycle muted by the introduction of a weak predator is also discussed.

6. Empirical results tend to show that omnivory is ubiquitous, although if the results of Pimm and Lawton (1978) are correct, it may tend to be everywhere but almost always weak. Recall that Pimm removed weak links of the diet in his food webs and found little evidence for omnivory. Further, where experiments have been conducted, omnivory has been a stabilizing force. However, experimental tests have not varied interaction strength to reveal the contingent stabilizing results theory predicts.

CHAPTER EIGHT

More Modules

In a general analysis of networks, Milo et al. (2002) found that food webs had an excess of diamond food web modules with and without intraguild predation (figure 8.1). More recent work on whole food web data has found that these diamond and intraguild predation modules are clearly overrepresented relative to randomly constructed networks (Bascompte and Melian, 2005). As discussed in chapter 7, these whole system results agree even with studies done on strongly interacting subsystems, which also found a preponderance of apparent competition and diamond modules [reviewed in Menge (1995)]. In what follows I will proceed by first investigating the dynamical outcomes of generalists before considering the dynamical properties of these ubiquitous food web modules.

In previous chapters I have emphasized that interaction strengths and topology can be used to identify whether a structure is stabilized or destabilized. Frequently, I have argued that a well-placed weak- to moderate-strength interaction can mute a potentially excitable interaction. In this chapter, I will find the same thing, but I will also highlight the idea that a generalist species can be stabilized by variance alone, regardless of interaction strength. Finally, the notion of a generalist foraging motivates the fact that food webs have an underlying spatial component that has often been ignored. As population ecologists have found, this spatial aspect begins to highlight the critical role space and behavior play in governing the dynamics of food webs—a topic that I will return to frequently in the final chapters.

8.1 GENERALISTS AND FOOD WEB DYNAMICS

Consumers are one of the major building blocks of food webs. Each consumer often has the potential to access multiple prey. The exact pattern behind the number of prey per predator thus is a critical determinant of food web topology. Further, trade-offs start to play a potentially major role in this aspect of food web structure, as there is evidence that there is a cost to generalism

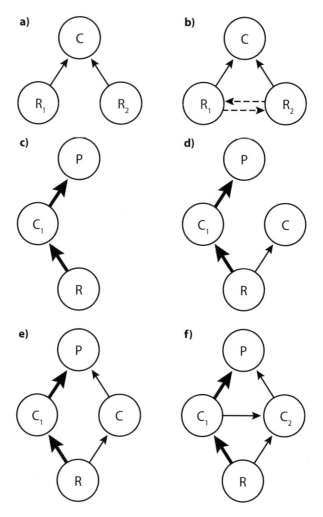

FIGURE 8.1. Some common modules. (a) Apparent competition or generalist; (b) with competition; (c) food chain; (d) with competition; (e) diamond module or generalist coupling multiple pathways; and (f) with intraguild predation.

such that more generalized species tends to have lower average consumption rates per prey item than more specialized consumers (Strickler, 1979; Drummund, 1983; Werner and Anholt, 1993). This trade-off suggests that differences in topology (i.e., low-connectance specialist webs versus reticulate generalist webs) likely manifest also as differences in coupling and interaction strength. Reticulate generalist-dominated webs would be expected to have lower average interaction strength than specialist-dominated food webs. I will turn to the implications of trade-offs later, but first I want to consider some theory for a consumer feeding on multiple prey with and without competition.

In order to understand the dynamical implications of generalism, I will explore a set of modules that contrast generalist consumer dynamics relative to the specialist case. Note that C-R theory, or the linear chains of C-R interactions that we discussed in chapter 7, define a theory for specialists. Food chain theory may also be envisioned as a theory about food webs that can be aggregated cleanly by trophic level [e.g., the assumption behind classic top-down trophic control theory (Oksanen et al., 1981)]. Our understanding of these specialist linear chain models is therefore important, as it acts as a control to measure the influence of generalism against.

First, I explore the role of generalism with and without exploitative competition (figure 8.1a and b), and to do this we require a model that allows coexistence of two competitors. A simple and informative model system under these assumptions can be written as

$$\begin{cases} \dfrac{dC}{dt} = -mC + ea_{CR_1}CR_1 + ea_{CR_2}CR_2, \\[2mm] \dfrac{dR_1}{dt} = rR_1(1 - (R_1 - \alpha_{12}R_2/K)) - a_{CR_1}cR_1, \\[2mm] \dfrac{dR_2}{dt} = rR_2(1 - (R_2 - \alpha_{21}R_2/K)) - a_{CR_2}CR_2, \end{cases} \qquad (8.1)$$

where all consumer parameters and functions have already been defined above, r_i is the rate of increase of species i, K is the carrying capacity of both resource species, and α_{ij} is the competition coefficient of species j on species i. This model with competition mimics the diamond-shaped food web module (Holt and Polis, 1997; McCann et al., 1998).

In the case of a generalist with noncompeting resources (i.e., $a_{12} = a_{21} = 0$), the generalist has the ability to unify the resource dynamics in the sense that high densities of a generalist manifest in simultaneously high consumptive pressures on the noncompeting resources. Figure 8.2a shows the response of the dominant eigenvalue to variation in the attack strength of the consumer on the alternative resource, R_2. Coupling into a new resource at no cost tends to make the interaction relatively less stable (figure 8.2a). This should not be surprising, at least when there is no competition between resources, because in essence generalism increases overall attack rates when the resources are synchronized. If we now allow the resources to compete in some capacity, the results also change.

Figure 8.2b shows the same experiment but now with competition. Competition was chosen such that R_1 is a stronger competitor than R_2

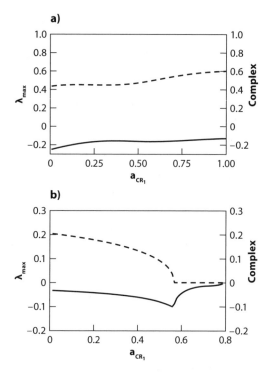

FIGURE 8.2. (a) No competition or switching, and complex eigenvalues tend to remain. (b) With competition, the now familiar pattern of stabilization and then destabilizion occurs. Stabilization reduces the complex signature in this case, as seen in previous chapters. Destabilization occurs here due to the exclusion of one of the resources. Note that destabilization can also include exciting the ever-strengthening C-R_1 interaction as well. In this latter case, another complex signature would increase during the destabilization phase.

($a_{21} = 0.80 > a_{12} = 0.6$). As we increase a_{CR_2}, the system is first stabilized and then destabilized. Thus, relatively weak interactions by the generalist on an alternative resource stabilize the module relative to the isolated module. The mechanisms behind the stabilization and destabilization phases are informative.

From previous chapters we expect that competition ought to act to shunt energy away from an otherwise strong pathway, here initially, the C-R_1 pathway (McCann et al., 1998). In this case competition acts to dissipate energy, and this means that the system should be stabilized. There are two possible reasons for the destabilization phase. One, the a_{CR_2} eventually gets strong, and so it can act to simply destabilize the system (McCann et al., 1998). Two, as this alternative C-R_2 interaction increases, this starts to strain the ability of these two resources to coexist. In figure 8.2b, coexistence occurs for relatively weak interaction strengths on R_2. Too much consumption on the inferior competitor pushs this species to local extinction.

The above arrangement of interaction strengths, which promotes coexistence, results from a simple ubiquitous trade-off. As long as the superior competitor is preferentially consumed, it is possible for the two competitors to coexist. One resource item survives via its abilities to compete for resources [e.g., the R* rule; Tilman (1982)], and the other resource survives because of its ability to avoid strong predation [i.e., the P* rule; Holt and Lawton (1994); Holt and Polis (1997)]. Thus, if high growth rates and strong competitive abilities tend to mean that you are also "edible," this condition ought to occur frequently in nature (Werner and Anholt, 1993). Weak-strong interactions in this generalist module ought to manifest frequently in nature, and so this stabilzing configuration should be extremely important in the real world.

The above mechanism (muting flux) has been played out in different forms throughout previous chapters, but this new module highlights the potential for a second mechanism that I have yet to consider. That is, the differential competing pathways coupled by the generalist consumer can potentially generate asynchronous dynamical responses in the two resources. This asynchrony has the ability to effectively take the wobble out of a consumer's dynamics because the consumer can average over the two prey items (McCann et al., 2000). This result might sound familiar, and it should. In a sense this is a generalization of the well-known single trophic level diversity-stability result often referred to as *averaging* (Tilman, 1996; Tilman et al., 1996, 1998). In this case, we are assuming that resource diversity averaging can stabilize a generalist consumer.

As an example of this, McCann and Rooney (2009) examined the dynamical response of a strong-weak parameterization of a Rosenzweig-MacArthur version of model system (8.1) after a perturbation was applied to both resources (see figure 8.3a for parameters). This model included density-dependent preference in foraging by the consumer such that consumers tended to move to patches with higher resource density [see McCann and Rooney (2009) for equations]. Thus, it included a rapid adaptive response of the top consumer to changes in resource density (I will return to the topic of behavioral responses in more detail in later chapters.) The perturbation employed was a modest but sustained increase in both of the resource-carrying capacities, K.

Both resources, therefore, initially increase in synchrony in response to the perturbation, as seen in region S in figure 8.3a. This synchronous early response, though, soon changes to asynchronous resources dynamics (region

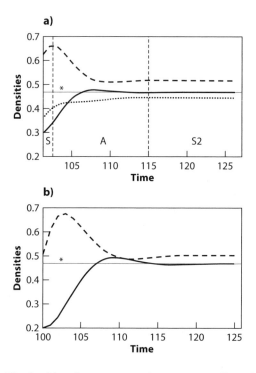

FIGURE 8.3. The densities of a consumer and two resources (R_1 and R_2) plotted after a perturbation at time 100. The horizontal line indicates the new equilibrium consumer density (*). (a) Asymmetric energy flux through two resources results in weak-strong interaction pathways (Parameters: $r = 1.0$; $K = 1.0$; $a_1 = 2.50$; $a_2 = 1.50$; $\alpha_{12} = 0.10$; $\alpha_{21} = 1.1$; $e = 1.0$; $m = 0.50$; $\omega = 0.50$). Post-perturbation dynamics starts with resources characterized by synchronous dynamics (region S) followed by asynchronous dynamics (region A) and then back to synchronous but stable dynamics (S2). (b) Symmetric flux through two resources results in the two resources being synchronized throughout the postperturbation period (Parameters: $r = 1.0$; $K = 1.0$; $a_1 = 2.0$; $a_2 = 2.0$; $\alpha_{12} = 0.60$; $\alpha_{21} = 0.6$; $e = 1.0$; $m = 0.50$; $\omega = 0.50$). Note that the greatest stability (quickest return time) occurs with asynchrony case (a) as indicated by the consumer density approaching equilibrium faster. Note also that in the completely synchronous case (b), the consumer, C, overshoots the new equilibrium (*) to a greater extent than in the asynchronous case (a).

A in figure 8.3a). The asynchrony occurs because any increase in total R is soon met with an increase in the generalist consumer, C. As C preferentially consumes the best competitor, R_1, then C ultimately begins to reduce R_1 as C increases. As R_1 declines, R_2 is suddenly freed from the competition, and so R_2 increases as R_1 decreases (region A in figure 8.3a). R_2 increases because the negative impact of increased consumption by C (which is weak on R_2) is smaller than the indirect gains from relaxation of the strong competitive pathway (C-R_1-R_2). This is a classic example of indirect interactions outweighing a direct interaction. If C is reduced instead of increased, a similar set of reasoning drives asynchronous resources dynamics.

The key to this asynchronous response and the rapid return to equilibrium in figure 8.3a is the differential pathways. To highlight this, McCann and Rooney (2009) chose parameters that created two symmetric pathways with moderate interaction strengths. This symmetric set of interactions was chosen in such a way that C still received approximately the same amount of potential production from both resources in order to isolate the change in pathway strength alone. Figure 8.3b shows that in this case the two resources stay synchronized permanently after the perturbation. Consistent with recent theory (McCann et al., 2000; Rooney et al., 2006), the weak-strong pathway scenario returns to the equilibrium (line marked * in figure 5a and b) more rapidly than the symmetric model (see figure 8.3 legend for details). Further, the symmetric pathway overshoots the new equilibrium line to a greater extent than the fast-slow pathway (i.e., the C trajectory in figure 8.3a barely rises above the equilibrium line, while the trajectory in figure 8.3b rises considerably more above the equilibrium line). The fast-slow pathway, therefore, drives asynchronous resource dynamics. In turn, the consumer averages over the asynchronized resource variance, thus increasing its return to equilibrium and simultaneously reducing overshoot dynamics.

This result can be readily generalized to abiotically driven variability. To highlight this idea, McCann and Rooney (2009) employed a simple stochastic model to show that variability in resources in space can strongly stabilize consumer dynamics (i.e., bound their dynamics away from low densities) as long as that variability in space is not synchronized. The stochastic model, not presented here, made the simplifying assumption that the consumer has absolutely no effect on the resource densities. Rather, resource densities were modulated purely by abiotic conditions. This assumption makes analysis easy and allows one to ask if a noisy environment, in and of itself, can be stabilizing. To examine this McCann and Rooney (2009) considered two endpoints, one where the underlying resources were periodically varying but perfectly synchronized, and another where the same resources were periodic but negatively correlated.

The synchronized resource forcing case not surprisingly produces bottom-up driven fluctuations of the consumer (figure 8.4a). Here, if we also simultaneously follow the food web configurations in the separate patches of this two-patch model, we find that the food web configuration remains the same in

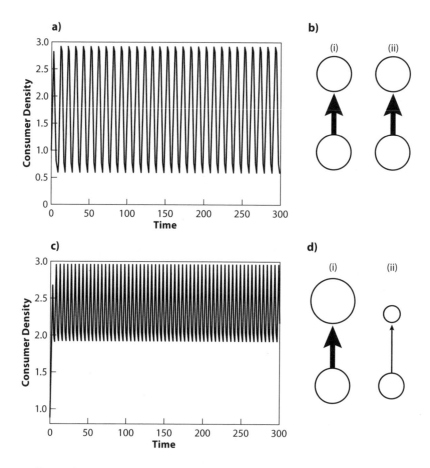

FIGURE 8.4. (a) Completely synchronous resources have wildly fluctuating consumer densities. (b) Completely synchronous C-R interactions with flux rates (arrows) in space at a given time when Rs are relatively high. Interactions are symmetric in space such that there are equal numbers of consumers and resources as well as identical consumption rates in both patches. (c) Completely asynchronous resource dynamics result in much more bounded consumer fluctuations. (d) Consumer resource interaction in each patch at a given time (relatively high resource densities in patch 1). Interactions are asymmetric in space such that patch 1 is a two-trophic-level system, and patch 2 more closely resembles a one-trophic-level system. Parameters: $a = 1.0$; $e = 1.0$; $m = 0.50$; $\omega = 0.5$.

both patches (figure 8.4b). This occurs despite the fact that the consumer has the ability to adapt. The reasoning for this result is trivial—the synchronization of resources in space eliminates variability in space, and so the system, in essence, acts like a single-patch model. Hence, the food web is static in space under synchronous or homogenized resource dynamics. Perhaps not surprisingly, then, large unstable resource fluctuations transmit through the consumer as large, unstable consumer fluctuations (figure 8.4a).

The perfectly asynchronous endpoint, though, gives the consumer heterogeneity in space and time. Figure 8.4c displays the dynamics of the same consumer foraging on resources that are completely asynchronized in space. The consumer still fluctuates, but it is easy to see that fluctuations in consumers are strongly muted relative to the individual resource fluctuations. A more bounded solution means a more persistent assemblage. Notice here, though, that the dynamics never are completely stabilized. In fact, what tends to happen is that the minima of the plot in figure 8.4c are lifted and the maxima are reduced when compared to figure 8.4a. This makes sense in light of the preference-based foraging decision chosen by McCann and Rooney (2009). Organisms tend to reside most of the time in higher-density patches and so elevate their minima relative to the synchronized case. Further, because of imperfect foraging decisions, some organisms remain in the low-density patch and so reduce the maxima relative to the synchronized case (figure 8.4a).

Changing the precise foraging decision does not qualitatively modify this result unless an organism forages severely suboptimally. For example, even a random foraging decision would allow the consumer to average over the two patches and so be effectively quite stabilizing. On the other hand, if all consumers dispersed from high resource density patches to low resource density patches, this would drive the consumer population to consistently low densities. Clearly, such suboptimal foraging would lower the chances that such a consumer can persist. Also, delays in the foraging response can counter this result. Abrams (1999), for example, has found that adaptive but slow responses to changing resources can instead destabilize the system. Nonetheless, as long as the switch is rapid relative to the scale of the prey dynamics, the result is generally stabilizing.

In this latter case of a heterogeneous world, the food web adapts on each patch in the two-patch landscape such that the food web is different in both patches. If R_1 is high and R_2 is low, then patch 1 has a strong C-R interaction and patch 2 has relatively few consumers feeding on R_2 (figure 8.4d). Patch 2 is effectively acting like a single trophic level (figure 8.4d). The asynchronization of resources in space and time allows the adaptability of this greatly simplified network to promote more stable consumer dynamics. The food web on each patch expands like an accordion, moving back and forth according to local resource densities. When resource densities are high, a *bird feeder effect* occurs, attracting consumers, and when resource densities are low, consumers leave the patch for the alternative higher-density patch. There is a cost to such movement, omitted from the model, but given even modest spatial variability

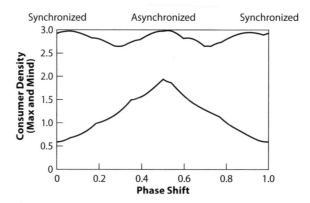

FIGURE 8.5. Consumer density, maxima and minima, after a transient of 200 time units as a function of the phase shift (period) between the two resources. Resources are completely synchronized at 0 and 1 full period and completely asynchronized at a 0.5-period phase shift. The greatest stability occurs with the greatest asynchrony. Parameters: $a = 1.0$; $e = 1.0$; $m = 0.50$; $\omega = 0.5$.

and finite costs, the stabilizing result robustly remains. The result is a highly dynamical food web but a persistent food web on the landscape.

To more generally explore the role of asynchrony further, we plotted local maxima and local minima of consumer density as a function of the phase shift between resource dynamics (i.e., from a zero-period shift to a full-period shift). Clearly, the result discussed above is general (figure 8.5). As the resource dynamics become less synchronized, the consumer dynamics approach a more bounded solution. The completely asynchronized case in figure 8.5 occurs at a phase shift equivalent to a half-period. Any amount of spatial resource variability, therefore, promotes the stability of such an adaptive mobile consumer (figure 8.5).

8.2   THE DIAMOND AND THE INTRAGUILD PREDATOR

As discussed above briefly, there is evidence for organismal trade-offs that play a role in governing the diamond module because trade-offs frequently manifest around parameters directly or indirectly involved with the coupling strength (e.g., a high tolerance to predation tends to drive a life history with lower growth rates). As such, life history trade-offs often govern the flux of material in any given set of interactions. I have argued above that such trade-offs play a major role in the flux of matter, the organization of interaction strengths, and ultimately in the stability of communities. It is not surprising that the diamond module and the intraguild predation module are intimately reliant on such trade-offs.

As an example, some organisms have adapted to grow rapidly, allocating energy to somatic and reproductive efforts, with little to no energy allocated to costly physical structures that impede a predator's ability to consume them. Such a consumer can be expected to be an extraordinary competitor in many situations. Nonetheless, their ability to produce biomass also makes them readily edible to potential predators. Thus, this species is both productive and potentially strongly coupled by a predator—it is the precise recipe for excited nonequilibrium dynamics.

On the other hand, an organism that puts much energy into the development of defense structures (e.g., a porcupine) may also be expected to grow more slowly and be less competitive when not in the presence of predators. A predator, though, has little influence on this organism, and so in the presence of predators (which by the nature of the trade-off love to consume the fast-growing consumer) this same slow-growing species may become an competitor. The trade-off therefore mediates a number of the following important attributes to this module:

(i) competition-defense trade-offs create the weak interaction distribution necessary for muting strong potentially excitable C-R interactions, and

(ii) competition-defense trade-offs create the necessary conditions for "generating asynchrony," thus enabling the predator to average over its variable prey items.

As discussed in chapter 7, it is well known that the model food chain exhibits several dynamical behaviors (such as stable equilibria, cycles, chaos, and multiple attractors). In a simplified sense, the food chain (the interaction chain, P-C_1-R) is best understood by considering it as two coupled consumer resource subsystems: a consumer-resource interaction (i.e., the interaction between C_1 and R) and the top predator consumer interaction (i.e., the interaction between P and C_1). Again, if the food chain is constructed from two strong consumer-resource interactions, the food chain population dynamical behavior can be quite complex and variable.

This reduction of the whole system to coupled subsystems allows us to take a simplifying view of this otherwise staggering problem. If one can find a mechanism that tends to stabilize all the underlying oscillators, this ought to eliminate the occurrence of cyclic or chaotic dynamics in the full system. Similarly, such a mechanism that reduces the amplitude of the underlying

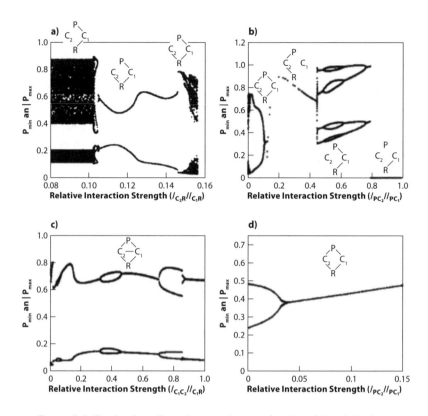

FIGURE 8.6. Food-web configurations are given as a function of the relative inter-action strengths. Whenever the configuration lacks an explicit link between a species and the rest of the connected web, this implies that the species cannot persist. (a) Exploitative competition. (b) Apparent competition. (c) Intraguild pre-dation. (d) The configuration used in (b), starting with a limit cycle solution. © 1998. Nature Publishing Group.

oscillators also ought to reduce the amplitude of the dynamics of the full system. Previous results in this book (e.g., the interaction strength corollary in chapter 5) suggest that this type of result may occur frequently.

While doing postdoctoral research at the University of California, Alan Hastings, Gary Huxel, and I sought to explore the role of small food webs from an interaction strength perspective (McCann et al., 1998). Toward this end, we performed theoretical experiments on a strong focal chain by adding interactions one at a time until the diamond module and the intraguild preda-tion module were both created. The results of these experiments are shown in figure 8.6a–d. This conceptual experiment included the additional assumption that trade-offs existed between a generalist predator and a specialist predator such that the generalist predator was not strongly capable of eating all prey items. The model assumed this by having a control parameter, ω, that toggled the strength in any one interaction. The parameter $1 - \omega$ was used for alternative prey such that as alternative prey consumption increased, the other prey item (ω) decreased.

The results of our 1998 paper were framed in a nonequilibrium context. In hindsight, I do not believe that they have to be so framed. The results, as shown in previous chapters, also can be demonstrated with equilibrium analysis using eigenvalues. Nonetheless, our result was that a stable consumer-resource interaction is required to dampen the dynamical behavior of a potentially excited interaction (and, hence, a potentially oscillatory interaction, at least in terms of its decay to equilibrium). At one level, when positioned properly within the food web (i.e., topology), a weak interaction can often act to decrease the relative interaction strength of a potentially strong interaction by increasing the loss term relative to the coupling term (sensu chapter 5).

As an example, figure 8.6a highlights a numerical experiment of adding exploitative competition, C_2, to a food chain model undergoing persistent chaotic dynamics. In this case, C_2 is competitively inferior to C_1, so its ability to persist is mediated by the selective predation of the top predator, P, on C_1. Figure 8.6a depicts the local minima and maxima attained for the top predator densities, P, in solutions across a range of C_2-R interaction strengths relative to C_1-R interaction strengths. Figure 8.6a also shows food web diagrams that depict the change in food web structure as the relative interaction strength changes value.

In this scenario we expect the exploitative competition mechanism to inhibit the oscillating C_1-R subsystem by deflecting energy away from the strong, potentially excited interaction. It does exactly this when C_2 can invade (at approximately 0.102 in figure 8.6a). As soon as C_2 can invade, the dynamics immediately begin to take on a much simpler periodic signal with population dynamics elevated further away from zero (i.e., further away from dangerously low densities). In this particular example, the system still does not reach an equilibrium solution over this range, as the muting potential of the added competitor simply is not capable of deflecting enough energy to cause period-doubling reversals all the way to an equilibrium value.

When the relative interaction strength increases still further, the attractor once again undergoes a period-doubling cascade to a more complex dynamical regime. This is not surprising—the new C_2-R interaction has become so strong that it no longer dampens the system but rather excites the system to chaos by adding a third oscillating subsystem to the food web. In terms of previous results, we have just walked through the familiar stabilization phase (inhibiting potentially excited interactions) followed by the destabilization phase (contributing excited interactions or similarly suppressing one of the species out of coexistence).

To study the full diamond module, we began with the same parameters as above except that we chose a starting point where all three interactions displayed oscillatory subsystem dynamics (P-C_1, C_1-R, and C_2-R). Then by constructing a link between the top predator, P, and the second intermediate consumer, C_2 (figure 8.6c), we were able to examine the role of interaction strength on the diamond module by toggling the interaction strength of P-C_2.

Figure 8.6b shows the local minima and maxima for the top predator, P, in solutions to equation (8.1) across a range of strengths of the P-C_2 interaction relative to strengths of the P-C_1 interaction. Again, one would expect the apparent competition mechanism to inhibit the C_2-R and the C_1-R oscillators. As relative interaction strength increases from zero, it immediately causes a period-doubling reversal, forcing simpler, more bounded limit cycle dynamics for relative strengths, approximately 0.03–0.10, and stable equilibrium dynamics for relative strengths, approximately 0.10–0.12. As before, increasing the strength of the interaction eventually destabilizes the system.

Despite the complexity of this system, which includes multiple attractors and numerous bifurcations, the qualitative result remains: relatively weak links, properly placed, simplify and bound the dynamics of food webs. Here, "properly placed" results from the underlying biologically expected trade-offs; this is important, as it is not just a mathematical trick. On the other hand, strong interactions coupled together are a recipe for chaos and/or species elimination.

Finally, a link was constructed such that C_1 fed on C_2 as well, creating intraguild predation within the food web. The parameters are the same as in the previous case, but with a relative interaction strength of approximately 0.01. For this final experiment we started off with a complex oscillatory dynamic (i.e., we again had three oscillators in the food web). Figure 8.6c shows the local minima and maxima for the top predator, P, in solutions to equation (8.1) across a range of the relative interaction strength. Here, we expect the apparent competition mechanism to inhibit the C_1-R subsystem, and we expect the food chain mechanism to inhibit the C_2-R subsystem. As a result of having two inhibitors and three potential oscillators, the dynamics never reach a locally stable equilibrium; they attain a well-bounded limit cycle solution (driven by the remaining oscillating P-C_1 subsystem).

In two of the cases above, weak links failed to beget stable local equilibria with all species present; at best we found well-bounded limit cycle solutions. This was largely a result of the choice of starting with chaotic dynamics and adding only one additional food web interaction at a time. To this end, we added another inhibiting weak interaction and found rapid local stabilization to an equilibrium solution. Figure 8.6d depicts solutions to the above system when we started with one oscillating solution from figure 8.6a. Here, we started with one oscillating subsystem (the P-C_1 subsystem) and then added the apparent competition mechanism, which inhibited the P-C_1 subsystem by extracting energy from this interaction. We rapidly get a locally stable solution for weak relative interaction strengths (at approximately 0.040 in figure 8.6d). Increasing the number of weak links relative to potentially oscillatory subsystems ought to tend to drive dynamics toward a stable equilibrium.

8.3 EMPIRICAL RESULTS

I now discuss some recent direct experimental tests of these results that come from fully replicated microcosm studies. In the remaining chapters when I begin to put all of the food web module theory together, I will explore these results more from the perspective of whole-system empirical patterns and field experiments. Both sets of results resonate, but the latter is more appropriately placed within a conceptual paradigm that addresses the dynamics of whole systems.

Rip et al. (2010) sought to test the diamond module in a series of aquatic microcosm experiments. Although Luckinbill (1973) had weakened interaction strength directly in an aquatic predator-prey experiment, most experimental work had not focused on looking at the role of interaction strength. We chose a relatively edible alga (*Scenedesmus acutus*) and a less edible alga (*Chlorella vulgaris*) being fed on by a rotifer (*Brachionus calyciflorus*). This experiment therefore created a natural diamond module extracted from open pelagic aquatic ecosystems. The experiment looked at each C-R interaction in isolation and compared it to the fully coupled diamond module. The prediction from modular theory, then, is that the fully coupled system is more stable than the system with the strong C-R interaction alone. Rip et al. (2010) measured numerous aspects of stability (table 8.1) and generated time series for the three treatments (figure 8.7a–c).

The dynamics of the diamond module showed no clear periodic signal and were less variable than the strong interactions. Additionally, the two algal resources showed signs of asychnrony in the presence of the rotifer as predicted (figure 8.7d). These results agree with those for another recent aquatic microcosm study seeking to test the role of weak interactions (Jiang et al., 2009), although in this experiment there was no evidence of compensatory resource dynamics due to asymmetric interaction strengths.

TABLE 8.1. Estimates of Interaction Strength and Resulting Measures of Stability for Each Experimental Treatment[a,b]

	Strong Interaction (B. calciflorus with S. obliquus)	Weak Interaction (B. calciflorus with C. vulgaris)	Coupled Interactions (B. calciflorus with S. obliquus and C. vulgaris)
Interaction strength [ln (Δ B. calciflorus) $\times$ day^{-1}]	0.50 ± 0.04^a	0.34 ± 0.04^b	$0.43 \pm 0.03^{a,b}$
Temporal variability			
CV algal biovolume $\times$ ml^{-1} $\times$ 10^6	0.74 ± 0.03^a	0.23 ± 0.01^b	0.55 ± 0.07^c
CV B. calciflorus biovolume $\times$ ml^{-1} $\times$ 10^6	1.00 ± 0.06^a	0.80 ± 0.10^a	0.49 ± 0.07^b
Boundedness from zero			
Min algal biovolume $\times$ ml^{-1} $\times$ 10^6	10.02 ± 1.09^a	84.34 ± 10.29^b	17.54 ± 3.84^a
Min B. calciflorus biovolume $\times$ ml^{-1} $\times$ 10^6	0.82 ± 0.17^a	5.67 ± 1.97^b	5.80 ± 1.04^b
Periodicity			
Max acf for algae where $p < 0.05$	14 days − acf ($n = 4$)	20 days − acf ($n = 4$)	23 days − acf ($n = 2$)
Max acf for B. calciflorus where $p < 0.05$	10 days − acf ($n = 5$) 20 days + acf ($n = 4$)	17 days − acf ($n = 8$)	16 days − acf ($n = 1$)
Mathematical stability (average model eigenvalue)	-0.066 ± 0.011	-1.49 ± 0.17	-0.19 ± 0.047
Dynamical behavior	Damped oscillations ($n = 5$);	Damped oscillations ($n = 1$); stable equilibrium ($n = 4$)	Damped oscillations ($n = 3$); stable equilibrium ($n = 2$)

[a] Numbers are means $\pm$ standard error. From ANOVA (d.f. = 2, 13) with Tukey's HSD analysis ($p \leq 0.05$), values with the same superscript letters in a given row are statistically the same, and values with different superscript letters are statistically different.

[b] n, Number of replicate microcosms; acf, autocorrelation function.

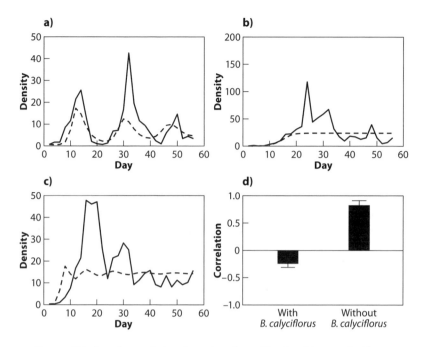

FIGURE 8.7. Sample population dynamics of a rotifer, *Brachionus calyciflorus*, for each experimental treatment. Presented are population density dynamics (solid line) and model dynamics (dotted line) for each of the three treatments. (a) The strong consumer-resource interaction. (b) The weak consumer-resource interaction. (c) The weak and strong consumer-resource interaction. (d) Correlation with and without the consumer. The correlation becomes negative in the presence of the consumer, with *B. calyciflorus*.

To this point we have looked at relatively simple modules and their connections. The results from even simpler modules, C-R interactions say, make sense in light of the higher-order subsystems. It remains to begin to consider the implications of connecting these modules to whole systems both theoretically and empirically. Importantly, some researchers have begun to address this aspect, as alluded to earlier in the book, and the results are fascinating.

In a recent innovative contribution that considers the mathematics of modules empirically, Kondoh (2008) noted that theory effectively argues there are two possible ways to yield a large, persistent system from underlying modules. First, the whole web is comprised of persistent subunits which add as building blocks to compose the larger web or a larger subweb. Second, the modules are arranged in a way that external modules support nonpersistent modules. In a sense, this is much like what we have discussed above. That is, interactions can be added such that they stabilize underlying unstable modules.

Kondoh (2008) used a Caribbean marine ecosystem and found evidence of both means for constructing more complex webs. Specifically, Kondoh showed that inherently stable intraguild predation modules were overrepresented in the Caribbean ecosystem. Second, from the other intraguild predation modules consisting of competitively inferior prey in isolation, Kondoh (2008) found that these modules tended to receive more advantage from external module interactions and in doing so enabled the persistence of these intraguild predation modules. Thus, Kondoh (2008) concluded that a whole food web "can be viewed as a set of interacting modules, non-randomly arranged to enhance the maintenance of biodiversity." This is a novel and powerful result, suggesting that the mathematics of natural modules may be understandable.

In a similar conceptual contribution, Bascompte et al. (2005) found that the interaction strengths in a large Caribbean marine food web were nonrandomly structured. They found that consecutive strong interactions within a food chain occurred less frequently than expected by chance. This suggests that strong and unstable interactions seem to be seldom coupled together in a food chain (recall that such consecutive couplings are a recipe for chaotic dynamics). Further, Bascompte et al. (2005) showed that when such strong couplings occurred in a food chain, they tended to be accompanied by omnivory more often than expected by chance. Thus, as seen in chapter 7, omnivory seems to materialize when strong overcompensatory potentials exist. This seems to be a strong argument for overcompensatory stabilization via omnivory in real webs.

8.4 SUMMARY

1. Empirical food web analyses have found that apparent competition and the diamond module with and without intraguild predation are ubiquitous (Menge, 1995; Milo et al., 2002; Bascompte and Melian, 2005).

2. If generalists are low in density (mathematically near a transcritical bifurcation), no-cost generalism on noncompeting resources/prey increases stability by elevating C densities; otherwise generalism on noncompeting resources tends to destabilize the generalist module (figure 8.2a).

3. If resources compete and the interactions are weak-strong in nature, this drives stabilization via two mechanisms: (i) weak interactions mute strong interactions, and (ii) the asynchronized resource/prey dynamics allow the predator to average over this variance in a way that promotes stability.

4. The ability of an organism to respond rapidly to variance (i.e., behavior) is critical to driving a rapid stabilizing response.

5. Results 2–4 above are from the diamond and intraguild modules. Weak pathways act to mute potentially wildly oscillatory pathways that result from strong interactions. Additionally, such different strength pathways have a tendency to create differential dynamical responses at common trophic levels (e.g., intermediate consumers respond asynchronously to changes in predator density).

6. The asymmetry of pathways may be common because of trade-offs between competitive ability (growth rates) and edibility.

7. Recent experimental tests have found that weak interactions in simple modules do appear to promote less variable population dynamics. Further, early analysis of whole webs shows that modules may add in ways that promote the persistence of whole webs (Bascompte et al., 2005; Kondoh, 2008).

PART 3
TOWARD WHOLE SYSTEMS

CHAPTER NINE

Coupling Modules in Space: A Landscape Theory

There are several possible approaches to addressing food webs at the landscape scale. The approach advanced so far in this book is to build toward whole food webs by understanding the addition of subwebs or modules. A second, more classic approach (discussed in the next chapter) is to study large matrices (i.e., whole communities) as championed by May and others (May, 1974a; Pimm, 1982). Still another strategy—and the one employed in this chapter—is to squint at the food web on the landscape, ignoring the details at first, in order to focus on the large-scale food web architecture. When we do this, at least for a number of ecosystem types, we find a relatively simple architecture that is deeply constrained by space. In essence, we stumble upon a large-scale module, or arguably a set of coupled food chain modules; however, this time when we deal with the module, we need to remain aware of the role of the spatial setting. Further, there is a hint from looking at the data behind this perspective that the root of this module may repeat across spatial scales and that this spatial module may tend to contain pathways with different interaction strengths.

9.1 VARIABILITY, SPACE, AND FOOD WEBS

To this point I have only briefly discussed the role of space when I considered generalists. Population ecology has demonstrated that space can interact with dynamics in fascinating and important ways [e.g., Comins et al. (1992); Hastings et al. (1997); Abrams (2007)], but it remains difficult to encompass the myriad patterns of spatial variability that occur across scales (Levin, 1992). When dealing with a complex system or problem, it can be of great advantage to look for ways to reduce complexity while retaining as much important information as possible. In this chapter I employ some empirical results to guide us in this manner.

As noted already, early food web empiricism tended to make the implicit assumption that food webs can be envisioned as static entities (Cohen, 1978; Pimm et al., 1991). This approach was consistent with classic theory that employed equilibrium assumptions (May, 1974a). These early simplifying assumptions are reasonable starting points, however, a number of empirical ecologists have since made cogent arguments for the importance of space and time in governing food web dynamics (Winemiller, 1990; Polis and Strong, 1996). Further, a growing body of complex adaptive systems theory has argued that some of the most fundamental aspects behind the persistence and functioning of any complex system may be its ability to respond to perturbations (Levin, 2003, 2005). While it is an exciting and innovative general theory for complex systems, it has proven somewhat elusive to concretely map these relatively abstract ideas to specific food web structures. This chapter is an attempt to piece together existing evidence to argue that variability in space, time, and food web structure, coupled with the ability of organisms to rapidly respond to variation, are critical to the maintenance of the food web and its functions.

Empirical and theoretical ecologists have now begun the difficult task of incorporating spatial and temporal aspects into food web ecology (Schoener, 1989; Winemiller, 1990; Holt and Loreau, 2002; Woodward and Hildrew, 2002; McCann et al., 2005). Although the early empirical results have been enticing (Eveleigh et al., 2007; Tylianakis et al., 2007), ecologists are in no position to empirically tackle this baroque problem using the modest number of food web studies that exist. Food web data rarely have been gathered with the required spatial and temporal gradients in mind (Martinez, 1991a). Additionally, documenting food webs in space and time is a monumental task. While traditional methods of gathering food web data remain necessary, a number of researchers are currently using body size and behavioral attributes of organisms to begin to project food web architecture across large spatial and temporal windows (Emmerson et al., 2004; Petchey et al., 2008; Rooney et al., 2008). This approach promises to make progress by accessing a large body of existing organismal data. This organismal level approach to food webs has the benefit of allowing ecologists to synthesize an enormous amount of existing empirical data on lower-level biological entities (e.g., individual species attributes and behavior) with the sparse amount of existing data capable of looking at large-scale empirical patterns in food web structure. Where this has been done, there have been some enticing consistencies across scales (Petchey et al., 2008; Rooney et al., 2008) suggesting that some unification of different subdisciplines may be possible within the food web framework.

9.2   INDIVIDUAL TRAITS AND A LANDSCAPE-SCALE MODULE

There is an established and well-documented research field in ecology that has sought out the ecological implications of body size [e.g., Peters (1983); Brown et al. (2004)]. This largely empirical literature provides a powerful base for understanding organismal traits and thus puts us in a position to make predictions about a food web structure that must adhere to these empirical relationships (Rooney et al., 2008). Ecologists have also produced an impressive body of research on individual foraging behavior (MacArthur and Pianka, 1966; Charnov, 1976) that can be placed within the food web framework. Taken together, body size and foraging behavior speak to food web structure in space and time.

It is commonly asserted that size correlates positively with trophic position (Jonsson et al., 2005; Arim et al., 2010). The data generally agree with this intuition, but occasionally the trend between size and trophic position is weak (Jennings et al., 2002a,b). One possible reason for some weak trends, though, may be that food chains of different-sized organisms from different habitats may be blended together in an analysis. A number of models that reproduce food web topology well amazingly do so using a single niche axis. This has been suggested to reflect the massive role of body size in structuring natural ecosystems (Cohen et al., 1985; Williams, 2000; Reuman and Cohen, 2004).

While trophic position often increases with body size, it is also simultaneously the case that larger organisms tend to be more mobile (Peters, 1983). Empirical research has found that the cost of transport is less in larger organisms (Peters, 1983). Thus, larger, higher trophic level organisms move more and are able to forage over larger spatial scales. These relationships among body size, trophic position, and spatial scale immediately set up a hierarchy of interactions in ecological space (see figure 9.1a). As a result of these empirical relationships, it is natural to suggest that small organisms, lower in the food web, are more isolated in space (McCann et al., 2005; Arim et al., 2010). As one moves up the trophic structure, we expect each new trophic level to progressively couple more spatially restricted organisms at the trophic level below them (figure 9.1a). Thus, in lower trophic levels, consumers effectively couple subwebs at a microhabitat scale, while higher trophic level consumers effectively couple subwebs at a macrohabitat scale. Finally, top predators can even couple distinct compartments or energy channels (Moore and Hunt, 1988). Such hierarchical structure has been recently discussed by network analysts for a number of complex systems (Clauset et al., 2008). This same hierarchy is

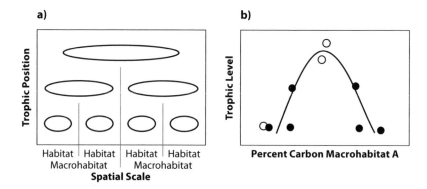

FIGURE 9.1. Figure depicting a hierarchically structured food web and the resulting fate of carbon originating from distinct habitats. Organisms low in the web
are isolated and so have a distinct habitat carbon signature, while higher-order
consumers progressively couple into more than one habitat and so possess mixed
carbon signatures. (a) The spatial scale of an average player on the trophic level
(ellipses) and the habitat it covers. (b) Where the carbon comes from in terms of
macrohabitat for the average consumer at a given trophic level. Dots represent
species, with the soild curve depicting the average trend across all species in the
food web.

strongly related to the food web structure long suggested by hierarchy theorists
(Kolasa and Pickett, 1989).

    If we follow the fate of carbon through the web depicted in figure 9.1 from
any distinct habitat, we expect organisms low in the web to get most of their
carbon from one isolated habitat (e.g., zooplankton get 100 percent from the
pelagic, and benthic invertebrates get 100 percent carbon from the benthos).
However, as we move up the food web, organisms begin to use carbon from
a wider range of habitats (e.g., a fish may get 50 percent from the pelagic
habitat, and 50 percent from the benthic habitat). If the hierarchical food web
prediction above is correct, we would expect a plot of the fate of carbon as a
function of trophic level to produce something akin to the hump-shaped plot in
figure 9.1b. The hump shape is effectively formed from the higher-order consumers coupling relatively isolated resources in space. If this is true, we predict
that behind nature's reticulate food webs lies a relatively simple spatially motivated module—the hump-shaped trophic structure depicted in figure 9.1b.
Further, if we define our habitats at different scales (microhabitat to macrohabitat to landscape), we may expect that this module repeats itself across a
variety of spatial scales whereby some fairly small consumers couple resources
from microhabitats and large consumers couple resources from macrohabitats, as fish do in the above hypothetical example. Vadeboncoeur et al. (2002)

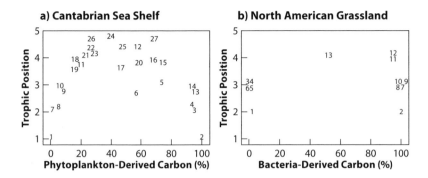

FIGURE 9.2. Two representative carbon channel graphs taken from the eight webs of Rooney et al. (2006). All eight webs are hump-shaped, suggesting progresive coupling into the two relatively isolated carbon sources as one moves up in trophic position. © 2006. Nature Publishing Group.

found exactly this result for a freshwater lake ecosystem, while in a related recent contribution, Arim et al. (2010) followed up on this food web structure to show that body size in killifish scaled significantly to the number of basal resources a consumer taps into. Here, larger killifish showed coupling into the phytoplankton, plant, detritus and terrestrial resource pathways, while smaller killifish were much more pathway-specialized.

To empirically assess this food web structure, Rooney et al. (2006) used eight of the most well-compiled webs that include estimates of flux rates between species. In all cases, the researchers followed the fate of carbon from distinct spatial habitats through the food web. In a sense, these eight webs represented systems with spatially distinct macrohabitats (e.g., benthic and pelagic) and systems potentially differentiated on finer microhabitat scales such as are found in soil webs (e.g., the dry soil fungal subweb and the moist soil bacterial subweb). Figure 9.2a and b displays the relationships across trophic levels for two of the eight webs investigated. In all cases, the predicted hump-shaped pattern occurred consistently. In figure 9.1b, organisms that derive relatively equal amounts of carbon from distinct habitats (e.g., the open symbols) are the mobile *food web couplers* of these otherwise relatively distinct habitats, and in this chapter we will find that these couplers play a critical role in mediating food web dynamics.

The above empirical results across ecosystems hint that this energy channelization may occur at multiple spatial scales. Figure 9.3 highlights an explicit example of this repeated structure in an aquatic ecosystem. Over a range of spatial scales, from within the water column (i.e., the pelagic subweb in figure 9.3a), to within the lake ecosystem (i.e., the pelagic-littoral

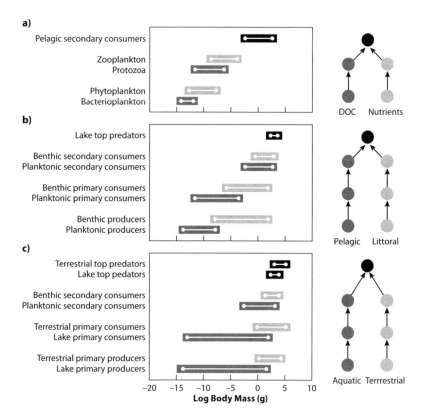

FIGURE 9.3. Three spatial scales of food webs from (a) within the pelagic, (b) between benthic and pelagic, and (c) between whole lakes and terrestrial ecosystems. Body size ranges from each pathway at a given trophic level accompany each figure. Notice that organisms from the slow pathways tend to be bigger and therefore may be expected to have lower turnover rates. Following the arguments laid out here, this also means weaker interactions and right-shifted consumer isoclines (see chapter 5). This large body size differentiation sets up a fast-strong and slow-weak pathway structure.

subweb in figure 9.3b), to between ecosystems (i.e., the terrestrial-aquatic food web in figure 9.3c), the pattern of relatively isolated pathways coupled by higher-order consumers repeats itself. Starting from the pelagic subweb (figure 9.3a), one finds that the dissolved organic carbon (DOC) pathway and the phytoplankton pathway are coupled by zooplankton (Fry and Sherr, 1984; Sherr and Sherr, 1991). As we move beyond this trophic level, we return to the whole lake web, which is coupled by mobile fish that move between littoral and pelagic habitats (Vadeboncoeur et al., 2002). And finally, at the landscape scale we find couplers, like birds and mammals, that integrate across terrestrial and aquatic prey (Hebert et al., 2008). Thus, at each spatial scale pathways connect via higher-order consumers—the hump-shaped trophic structure repeats itself across scales. This suggests nature may harbor a nested form of compartmentation.

It is also fascinating that in all the above cases one pathway tends to consistently have a smaller range of body sizes than the other pathway (e.g., DOC versus phytoplankton, pelagic versus littoral, aquatic versus terrestrial; figure 9.3). I will return later to this body size differentiation in order to predict another food web consequence of these different individual traits within the different pathways or subwebs.

Although the trophic position–carbon figures show distinct energy channels, this does not mean that competition between different compartments never occurs. For example, the benthic compartment of a water body can, and does, compete with the pelagic compartment. High densities of algae, for example, can shade out benthic production, and high densities of benthic production can limit essential nutrients to the pelagic compartment. Thus, even at fairly large scales there likely is some level of competition, especially when a system gets out of balance.

Some ecosystems do not follow the above patterns, as exemplified by the large, highly mobile herbivores of the Serengeti (Dobson et al., 2009). Nonetheless, there still remains an extremely mobile coupler; however, now some of the mobile herbivores act to couple resources in space, as well as coupling the more sedentary sit-and-wait predators at the top of the web (Dobson et al., 2009). This is a relatively subtle switch on the food web configurations in figure 9.2, and the Serengeti web, too, shows hierarchical arrangement as a function of body size, as discussed by Dobson et al. (2009). I would argue that the herbivores, as a result, may play an inordinate role in governing the dynamics and stability of these widespread ecosystems. Consistent with this, Fryxell et al. (2007) pointed out that the herding behavior of these wide-ranging organisms plays a major role in the stabilization of this ecosystem.

9.3   MOBILE ADAPTIVE CONSUMERS

Mobile organisms can couple networks of interactions in space and may do so at a variety of spatial scales (across microhabitats to macrohabitats). A consumer's ability to move between distinct resources that have different habitat demands puts them in a position to make foraging decisions between different habitat subwebs. Given that spatially distinct habitats vary in resource quantity, we expect the consumers to adopt behavior that tracks this resource variability (i.e., consumers move from a lower density prey habitat to a higher density prey habitat). The food web in the low resource density habitat therefore is expected to have a lower trophic position than the food web in the high resource density habitat—the food web, in a sense, expands and contracts vertically on the landscape like an accordion. This notion of the food web as an accordion has been called the bird feeder effect for the simple reason that organisms flock to areas of high resource densities as birds flock to a bird feeder in one's backyard (Eveleigh et al., 2007).

Here, I want to emphasize that I am looking at a food web's ability to adapt very rapidly (i.e., behaviorally). This rapid temporal response is important in considering whether a system is potentially stabilized by food web couplers or not. In many cases, fast responses to changing resource conditions are often stabilizing (McCann et al., 2005), while slower temporal responses can incur lags and may instead be destabilizing as emphasized by Abrams (1999, 2007). This remains an important empirical consideration. Hereafter, I consider that the couplers respond at fast time scales relative to the prey density population dynamics. In a sense, I am simply assuming that the predators arrive at a high-density patch when it is still a high-density patch and do not arrive with such a lag that a once high-density patch is at a low density. This latter case has been dubbed "antiswitching" (Abrams, 1999).

This ability to behaviorally adapt in space, though, requires us to think about the spatial scale of interacting species—a thorny issue looming over almost all problems in ecology [see the thorough account of Levin (1992)]. Following the lead of the general spatial food web theory of Holt and Loreau (2002), McCann et al. (2005) chose to create a spatially implicit food web model that allowed one to change the spatial scaling relationships of the resource and the consumer. In this sense, we can begin to theoretically suggest how space may play a role in interaction strength, dynamics, and stability. The idea effectively was to motivate a theory that took off from the empirical results discussed in the preceding section. de Koppel et al. (2005) independently wrote a paper that argued for an almost conceptually identical spatially scaled consumer-resource theory. In what follows, I first motivate some spatially implicit foraging equations that allow us to alter the scaling relationship of a highly mobile consumer with its resources before reviewing the food web dynamical implications of such behavior on the landscape.

9.3.1 SPATIALLY SCALED C-R INTERACTIONS

Recently, McCann et al. (2005) created a functional response that allowed us to theoretically alter the spatial scaling of the consumer-resource relationships

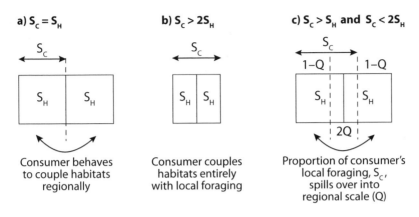

FIGURE 9.4. Three cases for the scaling of consumer-resource interactions where the consumer moves at larger spatial scales than the resource. (a) The consumer moves at a scale such that it perceives a single patch as well-mixed but moves between patches. (b) The consumer moves at a larger scale such that it perceives the two patches as well mixed. (c) Intermediate case where the consumer perceives one patch and some fraction of the other, Q, as well mixed. © 2005. John Wiley & Sons.

in a food web. This was accomplished by defining the spatial foraging scale of the consumer, S_C, as the area over which the consumer forages such that it perceives resources as well mixed. This consumer spatial scale is then considered with respect to the spatial scale of the resource habitats, S_H (figure 9.4a–c). Thus, this study assumed the consumer spatial scale, S_C, is a fixed biological trait of the consumer, while the scale of the resource is defined by the habitat type. Three informative cases emerge.

9.3.1.1 *Local Consumer Foraging Scale = Resource Habitat Scale* $$(i.e., S_C = S_H)$$

In figure 9.4a each consumer necessarily must make a choice between different local resource habitats when foraging at the regional scale. Such a case is likely to occur for spatially expansive systems where resource habitats can be larger than the inherent localized foraging scale of the consumer. Under the assumption that consumers are mobile and move on rapid time scales relative to population growth rates, McCann et al. (2005) derived the following spatially implicit multispecies functional response from a spatially explicit model:

$$F_i(R_i, R_j) = \frac{W_i(R_i, R_j)aR_i}{1 + W_i(R_i, R_j)ahR_i + W_j(R_j, R_i)ahR_j}, \tag{9.1}$$

where $W_i(R_i, R_j)$ is a preference function for the consumer on resource R_i given an alternative resource at density R_j, and is defined here as

$$W_i(R_i, R_j) = \frac{W_{T,i}R_i}{W_{T,i}R_i + (1 - W_{T,i})R_j},$$ (9.2)

and $W_{T,i}$ is the relative preference for prey i when all prey are at equal densities. The subscript T defines the trophic level of the prey. This mathematical simplification allows us to explore a spatially implicit model approximation of mobile consumers foraging on resources fixed in their habitats. Note that this case is similar in spirit to existing multispecies functional responses that include preference (Chesson, 1983; McCann, 1998; Post et al., 2000).

9.3.1.2   Local Consumers Spatial Scale $\gg$ Resource Habitat Scale (i.e., $S_C = 2S_H$)

The opposite case exists when a predator forages at a local scale (S_C) that is equal to the regional scale of the resource habitats (i.e., $2S_H$; figure 9.4b). At this scale of foraging the consumer forages in a manner such that it perceives the two prey from the different habitats as well mixed. It turns out that this case returns to the classic multispecies functional response

$$F_i(R_i, R_j) = \frac{aR_i}{1 + ahR_i + ahR_j}.$$ (9.3)

It is well known that the classic functional response implicitly assumes the consumer is feeding on prey species that are well mixed.

9.3.1.3   The Intermediate Case (i.e., $S_C > S_H$ but $S_C < 2S_H$)

A final case exists in which the consumer perceives the resource habitats as neither completely isolated (i.e., $S_C S_H$; figure 9.4a) nor completely well mixed (i.e., $S_C > 2S_H$; figure 9.4b). In this case an area of alternative habitat exists, Q, where $Q = (S_C - S_H)/S_H$, that is perceived by the consumer to be well mixed with the consumer's current habitat, while an area $(1 - Q)$ exists that requires the consumer to switch patches in order to gain access to the resources in that portion of the patch (figure 9.4c). Again, with these assumption in tow one can derive a multispecies approximation to this case that combines the two functional forms above [i.e., equations (9.1) and (9.3)], giving the following:

$$F_i(R_i, R_j) = \frac{S_i(R_i, R_j)aR_i}{1 + S_i(R_i, R_j)ahR_i + S_j(R_j, R_i)ahR_j},$$ (9.4)

where $S_i(R_i, R_j)$ is defined as

$$S_i(R_i, R_j) = Q + (1 - Q)W_j. \tag{9.5}$$

The key to these reductions is that we have effectively found a C-R interaction scaling parameter, Q, that allows us to change the relative spatial scale of the C-R interactions in a food web. When $Q = 1$, then equations (9.4) and (9.5) return to the classic functional response [equation (9.3)], while when $Q = 0$, the consumer must make a choice between foraging in the different habitats [equation (9.1)]. Thus, varying Q from 0 to 1 allows us to theoretically explore the consequences of compressing the spatial scale of consumer-resource interactions on food web dynamics.

9.4 FOOD WEBS IN SPACE

Figure 9.5a highlights a theoretical experiment that manipulates food web complexity in a spatially expansive world by sequentially adding higher-order consumers. In a sense, each addition of a higher-order consumer simultaneously increases the spatial extent of the focal food web because these consumers are increasingly more mobile. Throughout these numerical experiments the stability was assessed by following the local maxima and minima of the dynamics for the consumer species, C_1. The results that follow are insensitive to which species one follows.

For each new mobile generalist consumer addition, we look at the influence of the preference, w_i, on the dynamics. Each successive experiment starts by assuming a strong preference ($w_i = 1.0$) for one of the potential resources and then decreases this preference until there is equal preference for both resources ($w_i = 0.50$; the case of equal preference is equivalent to no preference). In order to test the influence of structure alone, each additional coupling is a mirror image of the existing isolated food web. Upon inspection of figure 9.5a, one can see that weak (i.e., $w_i = 0.95$) to intermediate couplings drive increased stability in all three cases. By increased stability, I am referring to the fact that the dynamics are less variable for weak to coupling strengths (figure 9.5a). This result resonates with the results of previous chapters.

In all cases, for equal preference (i.e., $w_i = 0.50$) the system tends to become unstable again. This is not surprising once one recognizes that equal preference by the consumer drives the resources in the underlying habitats to vary in phase. In a sense the equal preference eliminates any food web structure as the two resources begin to behave as one unified resource dynamically (the consumer synchronizes the resource dynamics). Each time a higher-order spatial coupler is added, the experiment proceeds

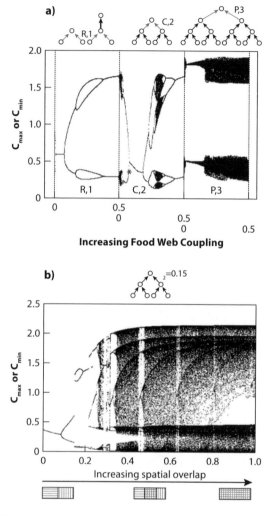

FIGURE 9.5. (a) Numerical bifurcation experiment of sequentially adding on higher, more mobile predators, which couple more and more into space. With each consumer addition McCann et al. (2005) started with $w = 0.5$ in order to destabilize the system before adding on a new trophic level. With each addition preference is varied and stabilization occurs with weak to moderate couplings. (b) Numerical bifurcation experiment of constraining spatial scale. Dynamics are greatly destabilizing by reducing the spatial scale of the ecosystem. © 2005. John Wiley & Sons.

by first purposely returning to greater instability (i.e., $w_i = 0.50$). The fact that this single mechanism is capable of stabilizing the successively increased production that accompanies these couplings shows the potent stabilizing force of this underlying mechanism.

This result is an extension of spatial predator-prey theory that has also found that, in general, highly mobile predators stabilize prey dynamics, especially when prey are asynchronous in space (de Roos et al., 1998; Holt and Loreau, 2002; Maser et al., 2007). The key to this result is that variable, and out-of-phase, resource responses mediate consumer stability via the consumer's ability to respond rapidly relative to the prey population dynamics. This multi-trophic level concept is a relatively straightforward extension of the statistical averaging effect and the negative covariance effect (Tilman, 1996).

Having considered a spatially expansive situation, it is natural to next consider the case where space may become constrained relative to the movement of the higher-order consumers. The conceptual experiment is to allow all habitats to decrease equally in size while retaining the spatial scale of the consumer (a fixed trait). This reduction in ecosystem size eventually leads to the case where the consumer therefore sees all resource habitats as well mixed (i.e., $Q = 1$). Figure 9.5b shows this experiment for a given constant value of w_2 ($w_2 = 0.85$; figure 9.5a). Reducing space has the consequence of increasing the consumer's influence on the resources by reducing the consumer's switching potential and increasing the mean interaction strengths. We see from figure 9.5b that the effect of this simplification of the spatial structure is to greatly destabilize the food web dynamics. Specifically, as resource overlap increases, the dynamics become increasingly more variable. Spatial constraints, therefore, increase the overall coupling and interaction strength on the resources and dramatically reduce the stabilizing potential of consumers integrating over separate and variable habitats (figure 9.5b). Consistent with this destabilization, spatially constrained food webs also tend to produce non-Eltonian biomass pyramids relative to more spatially unconstrained food webs.

There exist several possible mechanisms capable of alleviating such destabilizing forces imposed by the mobile consumer in a compressed space. Refuges impose limits on the extent to which a predator with no switching possibilities is able to suppress a given prey, while omnivory ought to accompany food web instability and the cascading densities in the trophic structure that the above instabilities drive. As an example, if a predator's prey is scarce but the prey's prey is abundant, the evolutionary or behavioral decision to short-circuit the food chain becomes enticing. Omnivory, at least moderate amounts of omnivory, has the potential to stabilize these spatially compressed webs. In this situation, omnivory provides a switch point and so acts as "ecological space." Thus we expect spatially constrained food webs to show a shorter food chain length (i.e., greater omnivory) and an increased amount of refugia behavior. Other mechanisms may also exist.

This general theory suggests that many of the attributes that we see in natural food webs may be the result of spatial compression (e.g., instabilities generating cascades and inverted biomass pyramids) and the responses that organisms make to it (increased omnivory and increased adaptations to refuges). Perhaps many of the patterns of distribution and refuging behavior are adaptations that emerge rapidly under spatial constraint and thus may be far less pronounced in a spatially unconstrained food web like the open sea. To my knowledge, this empirical prediction has never been investigated.

9.4.1 EMPIRICAL EVIDENCE FOR SPATIAL CONSTRAINTS

A natural place to consider the role of spatial constraints on food webs is in lakes, islands, and fragmented habitats. Lakes are an informative countercase, as they constrain space significantly relative to the enormous spatial scale over which fish move in the ocean.

McCann et al. (2005) considered the general theory developed above using data gathered for lake trout food webs across a range of lake size. Lake trout are generalist predators that are usually the top piscivores in deep Canadian lakes, and researchers have compared their size and trophic ecology across a whole series of lakes ranging in size from a few hectares to $>50,000\,km^2$. The lake trout trophic position (TP) in large lakes (TP $= 4.5$) is similar to that of free-ranging ocean salmon, but in agreement with the above theory, trophic position decreases sharply with lake size to around 3.5 in lakes as small as a few square kilometers (figure 9.6). The decrease in lake trout TP results from increased omnivory; that is, lake trout consume more benthos and zooplankton in smaller lakes, and fish make up a smaller proportion of their diet (Zanden and Rasmussen, 1999; Post et al., 2000).

Another factor contributing to shorter food chains in smaller lakes is the general decrease in the species richness of the prey community. As lake trout consume more small prey in smaller lakes, energetic constraints reduce their trophic efficiency, and this has been argued to lower growth efficiency, slow down growth rates, and reduce adult size (Sherwood et al., 2002a,b). These results are consistent with the general theory presented above and suggest that lake trout movement begins to spatially homogenize food webs below $100\,km^2$, driving a greater need for alternative stabilizing mechanisms (figure 9.6). It seems unlikely that the lake trout do not reach a higher trophic position simply because of energetic constraints (i.e., productivity), as the density of these lake trout in small lakes, where studied, has been found to be higher than that of large lakes (K. S. McCann, personal observation). Additionally, the loss of lake trout prey forage fish diversity is consistent with the destabilizing impact lake trout historically put on such a spatially

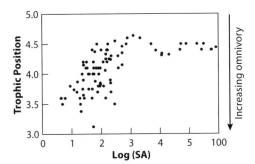

FIGURE 9.6. Plot of trophic position versus lake size (square kilometers). Each dot represents a separate lake. The trophic position decreases with decreased lake size, indicating increased omnivory. Figure modified from McCann et al. (2005). © 2005. John Wiley & Sons.

constrained lake. Finally, the data on lake trout trophic position appear more variable in lakes less than 100 km^2. Again, this is consistent with the idea that mobile predators can destabilize spatially constrained food webs.

When warm-water piscivores such as bass invade lakes, they tend to dramatically reduce the abundance and species diversity of forage fish species. This in turn appears to dramatically alter the lake trout's utilization of the littoral pathway (Zanden and Rasmussen, 1999). However, this effect, while clearly demonstrable for small lakes (<5 km^2), has so far not been detected in large lakes. In small lakes the system may be so compressed that little can be done to unfold more ecological space (habitat dimensions). Clearly, fish species that utilize riverine habitats in conjunction with lakes, or species whose temperature requirements more closely matched the range seen in a given lake, would be expected to suffer less spatial compression than a glacial relict like lake trout.

Since real-world food webs on spatially constrained habitat islands do not usually collapse down to a few species per trophic level, I expect that refuges must be present at all trophic levels. While many lakes and island ecosystems are much smaller than the basic ranging tendencies characteristic of the size of animals that occupy them, there usually is a structure of the environment that makes part of the system less habitable than others to most species. Most Canadian lakes, for example, have been invaded by at least a few warm-water species (e.g., pike, minnow, and bass), and as a result the coldwater fish have generally become confined to the pelagic zone (Zanden and Rasmussen, 1999). Thus, the ecological structure present in the environment provides refuges that can stabilize the food web in a spatially constrained world. Although refuges provided by ecosystem structure might appear to alleviate spatial constraints, the lake trout data show that there is still clear evidence of spatial constraints acting on lake ecosystems.

A compelling piece of evidence supporting the idea that even the largest lakes impose spatial constraints on their food webs comes from the dramatic top-down pressure that marine lamprey (*Petromyzon marinus*) can exert on lake salmonids when they obtain access to the Great Lakes, which was a major contributor to extirpations (Jensen, 1994). By contrast, in the open sea, where the lamprey feed on a similar salmonid food base, they seem to have a rather minimal impact; in fact, runs of returning anadromous salmon and trout usually have only a few percent of their individuals affected by these parasite/predators (Joseph Rasmussen, personal observations). Other "island" systems have been found to show similar trends. Post et al. (2000) and Post (2002), for example, have documented a similar trophic position decrease with decreased lake size for a range of temperate fish. Similar to the lake trout case discussed here, they suggest that the trophic position decrease was likely due to reduced prey diversity and increased omnivory. Further, Ebenhard (1988) revisited Elton's small island hypothesis by performing a comprehensive survey of the impacts of invasive mammals and birds as a function of island size. Ebenhard (1988) found that these higher trophic level organisms tended to have the greatest impact on islands versus land bridges and continental ecosystems, suggesting that small ecosystems are the most vulnerable to higher-order predators. Ebenhard (1988) also found that omnivorous predators tended to have less impact than strictly predatory species.

Both mathematical modeling and microcosm studies suggest that complete spatial overlap between consumers and resources readily leads to instabilities and strong runaway consumption unless (Elton, 1958) damped by refuging or omnivory. While many empirical food web researchers have considered such analyses to be simplistic small-scale artifacts because real complex food webs appear to be much more stable than models indicate, it can be argued that the vulnerability of small ecosystems to disruptions when invaded by large predators is evidence that such models are in fact valid under constrained conditions. Further, it can also be argued that lakes constitute highly constrained aquatic ecosystems compared to those of the sea and that small lakes ought to be extremely sensitive to dynamical instability and top-down cascades (i.e., non-Eltonian biomass pyramids).

9.5 ASYMMETRIC FLUX RATES THROUGH FOOD WEBS

Above I have motivated a food web structure based on average organismal traits and pointed out that we expect increased interaction strengths in spatially constrained food webs. Food webs tend to be increasingly coupled by higher-order trophic level species. I have also argued that this pattern of coupling may occur at small scales (small consumers couple across microhabitats) and larger scales (large consumers may couple across whole ecosystems), and further, that coupling—at least in spatially expansive systems—ought to be a potent stabilizing mechanism.

Yet, besides pointing out that overall interaction strength ought to reduce on average with increased ecosystem size, I have said little about the structure of interaction strengths in food webs. It is worthwhile returning to organismal traits, such as body size, to see if there are allometric reasons that we may expect clear differences in flux rates through different pathways. As previously discussed, compartments or channels exist within the pelagic zone (figure 9.3a), within the whole lake (figure 9.3b), and between entire ecosystems (figure 9.3c). Figure 9.3a–c also shows that a clear pattern exists across scales: On average, one channel or compartment tends to be smaller in mean body size consistently over all trophic levels (figure 9.3a–c).

This again puts us in a position to consider the implications of different-sized organisms by turning to the extensive empirical literature on body size and organismal traits. Specifically, the allometry literature has pointed out that small organisms tend to have high turnover rates, while large organisms have lower turnover rates (Peters, 1983). In a sense, the range in body size forms a strong-weak metabolic continuum whereby small organisms have a heightened cost of living [basal metabolic rate (BMR)/weight is relatively high] and require high consumption and growth rates to match this high energetic cost. The result of such simple mathematical relationships is necessarily, on average, a higher turnover rate in smaller organisms (Brown et al., 2004). These simple empirical and theoretically reasonable relationships imply that pathways with different-sized organisms also ought to have different rates of turnover. Therefore, pathways with small organisms have higher turnover rates (strong or fast pathways), and larger organisms have lower turnover rates (weak or slow pathways).

Recent research has found this to be precisely the case for a number of aquatic and soil food webs (Rooney et al., 2006). The above body size relationships across scales, coupled with empirical body size relationships, further suggest that we have such strong-weak pathways across the range of food web scaling (i.e., from within food subwebs at a microhabitat scale to food webs that span entire ecosystems). This prediction has yet to be fully explored, but early empirical analysis, and the consistent difference in body size between subwebs, suggest it may occur frequently in ecological networks.

Let us now place these empirical ideas within the consumer-resource framework developed in this book. For example, what do these turnover rates mean for average isocline relationships? Are strong flux rates more likely to be governed by left-shifted consumer isoclines that are inherently less stable (see chapter 5)? Allometry suggests that consumption and loss rates scale to the same power (per unit weight these rates scale to -0.25). Thus, there is little reason to suggest that higher turnover rates imply a less stable interaction. However, other relationships with body size suggest that smaller organisms, all else equal, might tend to be less stable with more left-shifted consumer isoclines. Both per capita growth rates and carrying capacity tend to scale inversely with body size such that small organisms are more productive. This immediately suggests that the relative position of the consumer isocline is more left-shifted for small organisms. Further, there tends to be less skeletal material with smaller organisms, meaning that the conversion of biomass per unit of biomass (e in C-R equations) in smaller organisms is likely higher on average. This is another argument for smaller organisms having greater flux rates, greater turnover, and more left-shifted consumer isoclines (i.e., less stable dynamics). Terrestrial organisms, for example, are not only slightly larger than aquatic organisms, on average (although there is quite a range of body sizes in both types), but also tend to have more dense skeletal structures because of the different physical demands of the terrestrial ecosystem. Terrestrial plants require substantial infrastructure to hold themselves vertically to capture light, while aquatic plants are often tiny plankton that are small and bouyant, minimizing sedimentation rates (and loss of sunlight). Land mammals also require more substantial skeletal structures than the large herbivores and predators of aquatic systems, fish, which have light cartilagenous skeletal systems. Thus, the evidence leans toward the idea that small body–size pathways ought to generate strong interaction strengths while large body–size pathways contribute weaker interactions.

Here, I have argued that food webs tend to show compartmentation of the sort discussed above (channels coupled by mobile consumers) but also tend to show differential rates of flux and therefore differential stability properties. The question remains, How does the coupling of these different-strength pathways influence stability? Results discussed in previous chapters suggest that this arrangement may have some strong stabilizing properties.

9.6 DYNAMICAL IMPLICATIONS ON THE LANDSCAPE

In chapter 8 I argued that a generalist consumer capable of behaving according to reasonable rules (i.e., tends to choose high-resource environments over low-resource environments) can be a stabilizing player. Effectively, these generalist consumers drive any food web within a habitat to expand in trophic levels (when prey/resources are abundant) and contract in the number of trophic levels (when prey/resources are low) in a manner that relieves resources from consumption pressure when densities are low and regulates unbounded resource growth when resource densities are high. The trophic structure may be expected to expand and contract like an accordion in response to resource fluctuations as long as resource dynamics are heterogeneous on the landscape. As such, generalists play a pivotal role in generating food web variability in topology within a habitat while buffering against extreme population dynamical variability through time (McCann and Rooney (2009)).

This food web response to prey densities in space and time has been called the *bird feeder hypothesis*, as discussed in the previous chapter (Eveleigh et al., 2007). It relies on the assumption, though, that different habitats do not experience synchronized resource production. If they do, these models would predict little food web variation and less stable dynamics. Thus, another fundamental element of stability are mechanisms that generate resource asynchrony. In previous chapters we discussed how weak-strong interactions by generalists can generate resource asynchrony, which the generalist top predators integrate over. Similarly, a consumer switching behavior can also generate resource asynchrony. Note that while some species may be synchronized across large spatial scales (Schwartz et al., 2002), the above stabilizing mechanism merely requires that other alternative resource species are not. Thus, the abundance of research on population synchronization is not necessarily counter to this result. Some recent evidence argues for an unusual amount of synchrony between species (Houlahan et al., 2007); however, this was a first step in addressing this issue, and it seems likely that at certain temporal and spatial scales both types of dynamics occur [e.g., see Vasseur et al. (2005)]. I will turn to an example in the empirical sections of both synchrony and asynchrony in a community.

The structure outlined here—generalist consumers that differentially couple resource habitats in space (i.e., weak and strong pathways)—fits the requirements for the theory motivated in the previous chapter. That is, the different strength pathways and the switching of responses readily generate asynchronous dynamics. Further, the mobility of the consumer organisms enables higher trophic levels to behaviorally adapt to variation in the landscape in a manner that is stabilizing. Note, as indicated above, that this is expected to occur in a spatially expansive ecosystem that promotes consumer/predator switching. Thus, this food web structure implies that a whole food web should readily display bird feeder effects and should also "generate resource asynchrony" in the face of major fluctuations in time and space.

I have outlined two major theoretical implications of the empirically determined food web architecture. First, the bird feeder effect allows food webs to adapt to variability in the landscape in such a way as to promote stability and persistence as long as resource patches on the landscape are sufficiently out of phase (i.e., not synchronized). Second, given that abiotic mechanisms may occasionally synchronize resources in space, the existence of fast-slow pathways coupled by mobile adaptive predators readily generates spatial asynchrony in the resources. Taken altogether, I have identified some simple empirically motivated food web structures that react to variability in a way that promotes balance in a variable world. Thus, food webs may be constructed such that they operate on the variability and even produce spatial variability, which in turn buffers adaptive consumers. I now review some empirical examples to show that some food webs do, indeed, display (1) the bird feeder effect, (2) the ability to internally generate asynchrony through fast-slow pathways, and (3) resource coupling of distinct habitats which appears to stabilize food webs.

9.7   EMPIRICAL EVIDENCE

9.7.1   Eveleigh's Balsam Fir Food Web and the Bird Feeder Effect
In a recent investigation, Eldon Eveleigh and colleagues (2007) were able to explicitly test the bird feeder effect within the context of a complex boreal insect food web. In an enormous field effort, Eveleigh et al. (2007) collected and analyzed 20 plot-years of interaction data (including more than 100,000 insect rearings). This extensive catalog enabled them to build a series of food web snapshots over a wide range of budworm densities. This research effort therefore had the data to catalog how the food web adapts as a function of resource variability. Additionally, the data were gathered from both heterogeneous forest stands (mixed hardwood and softwood) and homogenized forest stands (almost all balsam fir).

One expects the bird feeder effect within a variable spatial landscape of resources to make the food webs expand and contract according to local resource densities. Further, this accordion-like trophic response should be most dramatic in higher trophic level food web couplers. Consistent with this bird feeder effect, Eveleigh et al. (2007) found that the balsam fir food web expanded vertically when budworm densities were high and contracted when budworm densities were low. Higher-order generalist predators and parasitoids were responding in space to the outbreaks (figure 9.7). Curiously, the homogenized plot consistently showed weaker bird feeder effects (Eveleigh et al., 2007). Thus, these data suggest that human homogenization on the landscape

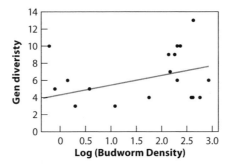

FIGURE 9.7. Response of balsam fir food webs to changing budworm densities. The number of secondary and tertiary generalist parasitoids increase with increasing budworm density. Thus, like a bird feeder effect, a cascade of higher-order parasitoids appears at high budworm densities.

may weaken this stabilizing mechanism. Consistent with this result, Eveleigh et al. (2007) found that parasitism rates were higher in the heterogeneous plots, suggesting budworm were being reduced by the onslaught of parasitic diversity.

To further experimentally test this result, Eveleigh and colleagues (2007) then created a localized budworm outbreak. Before the outbreak the web was depauperate and truncated, and shortly after the localized outbreak the web became more reticulated with a higher maximum trophic position. In all cases, it appeared as though secondary and tertiary generalist parasitoids were responsible for the expanding and contracting of the food web. This is an especially challenging test of some of the ideas of the theory, as this web is not structured by size in the general way suggested above. That is to say, the higher trophic level parasitoids tend to be smaller than their lower-level hosts. Nonetheless, there are still mechanisms that promote their movement on the larger landscape (e.g., chemical cues). There is an argument that the above result is merely optimal foraging in practice, and it most definitely is. Nonetheless, this foraging occurs at an extremely broad scale and does so with enormous dynamical implications.

9.7.2 PELAGIC SUBWEBS GENERATE ASYNCHRONOUS RESOURCE DYNAMICS

Limnologists have long noted the synchronization of plankton in early spring followed by the decoupling of edible and relatively inedible plankton in the summer [e.g., the phytoplankton ecology group (PEG) model of succession; Sommer et al. (1986)]. In the spring, nutrients sweep in off the landscape and create plentiful conditions for phytoplankton to flourish, the small edible

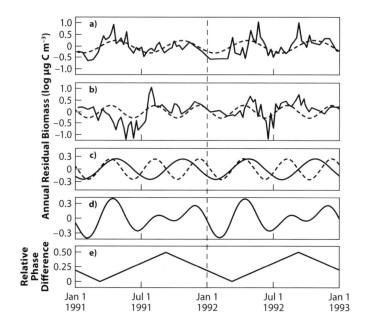

FIGURE 9.8. Asynchrony generation. Using statistical techniques to decompose time series into underlying periodicities, Vasseur et al. (2005) continuously estimated the phase shift in edible and less edible phytoplankton (0 is in phase or synchronous, 0.5 is out of phase or asynchronous). The plankton tend to respond by asynchronizing during the summer after a strong synchronizing event in the spring. (a) Edible plankton dynamics. (b) Less edible plankton. (c) Dominant periodicities of edible and less edible trajectories plotted together. (d) Summed dynamics. (e) Estimated phase difference. From Vasseur et al. (2005). © 2005. John Wiley & Sons.

plankton rising the most but other larger, less edible plankton also increasing (Vasseur et al., 2005). As a result, abiotic conditions readily promote a period of algal synchrony. Consistent with the simple diamond model presented in chapter 8, though, the herbivorous cladoceran density soon responds to these bountiful resource conditions and increases in density. This rise in herbivore density naturally intensifies herbivory, especially on the small, more edible phytoplankton Vasseur et al. (2005). Taken altogether, we have the perfect conditions for asynchrony generation in that we have high consumptive pressure on two competing resources and the existence of strong-weak pathways (i.e., a fast-growing edible algal functional group that is preferentially consumed and a slow-growing less edible algal functional group that is weakly consumed). Vasseur et al. (2005) examined 20 years of data for Lake Constance and noted this well-known seasonal succession with spring algal synchronization followed by algal asynchronization or compensation (figure 9.8).

To more thoroughly examine the mechanisms behind these coherent and compensatory dynamics, Vasseur et al. (2005) examined the 20 years of lake Constance plankton data by employing a number of sophisticated time series techniques that allowed them to continuously estimate the degree of synchrony between edible and less edible plankton in the planktonic subweb. Vasseur et al. (2005) then plotted both nutrient conditions (soluble reactive phosphorus) and cladoceran density versus degree of synchrony. The results showed a strong relationship such that low phosphorus and high cladoceran density correlated with asynchrony, while high nutrient conditions and low cladoceran density correlated with synchronous dynamics (figure 9.9). The asynchrony is precisely the result one would predict from the strong-weak food web diamond module (figure 9.9b). In a sense, the spring nutrient pulse ultimately heightens both herbivory and competition later in the season, and this simultaneously strong herbivory and competition generate resource asynchrony (figure 9.9b).

This example is consistent with asynchrony generation within a subweb, an empirical idea with a long history in limnological work (Sommer et al., 1986). It remains, though, to show that asynchrony generation can occur on larger spatial scales as well. A recent empirical analysis has found a range of bird switching responses operating on asynchronous invertebrate and fruit dynamics (Carnicer et al., 2008). The strong switching responses documented by these authors leave open the possibility that this is a larger-scale example of resource asynchrony generation, as the temporal period this unfolds over appears more rapid than seasonal dynamics alone.

The above theory, which relies so heavily on variability in space, also suggests that it will be fruitful to re-examine existing temporal and spatial data for a variety of resources to more fully understand both synchronous resource dynamics and asynchronous resource dynamics (Carnicer et al., 2008; Owen-Smith and Mills, 2008). We have concentrated on the role of weak-strong pathways, but other environmental mechanisms (like differential responses to abiotic conditions) may frequently be responsible for generating resource asynchrony. This aspect of asynchrony is firmly a part of the theory presented here. I have focused on asynchrony generation, as it is a less understood way a system responds in the face of synchronizing environmental conditions. Further empirical and experimental work determining patterns in synchrony/asynchrony within trophic levels, and what drives these patterns, will significantly aid the ideas presented here.

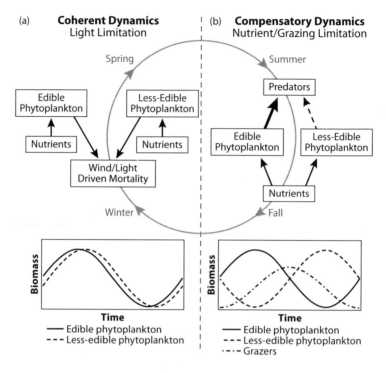

FIGURE 9.9. Spring forcing (a) and the weak-strong interactions that drive asynchrony in the summer (b). This latter idea is consistent with the PEG model of Sommer et al. (1986). From Vasseur et al. (2005). © 2005. John Wiley & Sons.

9.7.3 RESOURCE COUPLING AND STABILITY (ROMANUK ET AL., 2006)

The theory above suggests that generalist organisms capable of coupling into a diverse resource base can be enormously stabilizing. This idea is closely related to the 1955 paper by Robert MacArthur who argued that more pathways allowed a consumer/predator with options that buffer against the decline or loss of a single pathway. Recent theory, outlined above, has shown that this stabilization is highly dependent on the strength of the interactions and asynchronous resource dynamics. A diverse and synchronized resource base provides no such benefit to a consumer because when times are bad for a given resource, they are simultaneously bad for all resources—there is no reprieve from starvation for a consumer.

In a nice recent contribution, Romanuk et al. (2006) used stable isotopes to look at this problem. They followed up on MacArthur's idea by asking if organisms with a diverse resource base had more or less variable population dynamics. They measured the $\delta^{15}C$ of organisms in the food webs of tropical aquatic rock pools, and assumed that greater carbon variation meant that the consumer was capable of coupling into a greater range of distinct carbon resources. They found that increased carbon variation (i.e., increased coupling of distinct resources) correlated negatively with population variation. It is worth mentioning that this greater range in carbon variation does not mean that the consumer eats more resources; however, it does mean that the resource pathways span a wider range of niches. All else equal, these more divergent resources are therefore less likely to show synchronized responses to variable conditions. Thus, as predicted above, organisms capable of coupling into a wide range of resource pathways were more stable.

During this same analysis, though, Romanuk et al. (2006) found that variation in $\delta^{13}N$, which reveals feeding at divergent trophic levels, had no strong relationship with population variability. In fact, if anything, the data showed a weak positive trend such that more omnivory seemed to correlate with increased variability. This may indicate that omnivory is a response to increased resource population variability, as discussed briefly above and later in chapter 12.

9.8   SUMMARY

1. Food webs show a repeated fundamentally spatial food web module, generalist coupling resources with strong and weak pathways, at a variety of spatial scales (generalists coupling resources in different microhabitats to generalists coupling resources at the entire landscape scale).

2. These generalists couplers, and their ability to adapt to variation on the landscape can be potent stabilizers when embedded in a spatially expansive ecosystem. However, in a spatially constrained or restricted ecosystem, they become potent destabilizing influences.

3. Arguments based on body size and other organismal traits (digestibility) suggest that the pathways that generalists couple at a variety of scales tend to be distinctly strongly interacting pathways or distinctly weakly interacting pathways.

4. The above food web structure, placed within a landscape context, suggests that higher-order predators ought to respond en masse to variation in prey in a heterogeneous environment. This response, which expands the trophic structure locally, is referred to as the bird feeder effect.

5. Strong and weak pathways coupled by generalist predators readily generate resource/prey asynchrony even after a perturbation that initially synchronizes the resources. This is referred to as resource asynchrony generation.

CHAPTER TEN

Classic Food Web Theory

10.1 THE CLASSIC APPROACH

To this point I have emphasized the study of ecological modules and projected to larger systems by arguing that certain modular architectures recur across spatial scales, allowing us to look at stabilizing features at the landscape scale. This approach skips the myriad detailed interactions that actually exist in real webs. The classic approach, on the other hand, has tended to employ n-dimensional matrices that express the interactions among all n species of a whole community. This approach has the advantage of encapsulating an enormous amount of community information into a relatively simple matrix. Further, well-developed linear stability techniques can be used to ask if the community is stable or not, or even how stable the matrix is (i.e., the size of the dominant eigenvalue).

Because so much theory has derived from this matrix approach [e.g., May (1973); Pimm et al. (1991)] and communities truly are more complex than the modules we have looked at, it is important to ask whether results from the modular approach make sense in light of the whole-community matrix results. There appears to be a notion that these two theories are often at odds with each other [e.g., Gross et al. (2009)]. Here, I argue that both sets of results make sense in light of each other. In what follows, I first describe in more detail this classic whole-community approach and then introduce some simple mathematical techniques for understanding these results. I will end by showing that the ideas derived from these classic whole-system matrix approaches largely agree with the results of modular theory.

The classic community matrix approach assumes that any specific matrix represents a sample from a "statistical universe" of interaction strengths [sensu May (1974a)] for a given set of n species. There is no implied underlying structure to the interaction strengths, rather it is a random set of interaction strengths given a specified food web topology. Theoretically, researchers may wish to bound this statistical universe in some plausible way before sampling from the distribution of interaction strengths [e.g., Pimm (1979); Yodzis and

Innes (1992)] but, nonetheless, it remains randomly sampled from within these constraints. Once sampled, this theory has tended to ask if the n-species matrix produced is stable or not (i.e., whether the dominant eigenvalue is negative or not). Further, as the random sampling of interaction strengths from the assumed distribution is repeated, the percentage of stable webs (PSW) for a given food web structure or topology is calculated.

May (1974a) used this community matrix approach to determine if more diverse ecosystems (i.e., large n-species ecosystems) tended to have a greater chance of being locally stable than small n-species ecosystems (i.e., whether diverse systems have a higher PSW). Stuart Pimm and others followed similar procedures to ask if omnivorous or fully compartmented systems harbored a higher PSW than systems without omnivory or compartments (Pimm and Lawton, 1978; Pimm, 1979; Solow et al., 1999). Much less frequently, these same matrix approaches have asked if the magnitude of the maximum eigenvalue changed with or without a given structure. As an example of this latter use of community matrices, Pimm and Lawton (1978) found that although omnivory tended to be locally unstable (i.e., have a low PSW), when it was locally stable, it tended to be very stable (i.e., had a more negative dominant eigenvalue than an average food chain model).

It is necessary to recognize, then, that the matrix approach generally sought to understand if a given topology tended to produce stable or unstable webs. At this point in food web theory, there was yet little consideration that the answer may depend more subtly on interaction strengths [although May pointed out that interaction strength was fundamental (May, 1974a)]. In a sense this statistical universe approach asks if, on average, there are food web structures that have little likelihood of being stable. If there is little likelihoood, the assumption is that such a food web topology likely does not occur often in real food webs. Clearly, this is a reasonable question and, as it stands, suggests that omnivory ought to be rare given that Pimm and Lawton (1978) found that it rarely produced stable configurations.

This broadly posed question, though, potentially misses critical context-dependent modifications. As an example, while it may be true that most omnivorous systems are locally unstable (i.e., have a low PSW), it may also simultaneously be true that weakly omnivorous systems are strongly stable. In other words, the answer may depend on interaction strength, an answer that is hidden in the classic statistical universe approach. It is these interaction strength–dependent answers that modular theory seeks to find because the model experiments (e.g., bifurcation analysis) often examine the magnitude of stability changes across an important biological gradient (e.g., across gradients in interaction strength, strength of omnivorous interactions).

The question therefore becomes: Can we apply context-dependent aproaches to matrix theory? There seems to be little reason why not. To examine this I first review some matrix approaches.

10.2   MATRICES AND LOCAL STABILITY

Recall the interaction strength matrix of Robert May (i.e., the Jacobian) introduced in chapter 5:

$$\begin{bmatrix} a_{11} & a_{12} & \dots & a_{1n} \\ \dots & \dots & \dots & \dots \\ a_{n1} & a_{n2} & \dots & a_{nn} \end{bmatrix}, \tag{10.1}$$

where the subscript ij refers to the influence of species j on species i. More specifically, it refers to the rate of change in the population's growth rate $[dN_i/dt = F_i(N_i, N_j)]$ relative to the rate of change in another species' growth rate (mathematically, this is defined as $a_{ij} = \partial F_i/\partial N_j$ and is defined at equilibrium).

Further, recall that coupling terms and loss terms govern the fate of energy or material flux through any C-R interaction. The loss terms encompass both linear and self-damping losses. I argued in chapter 5 that the dynamical response of the basic C-R interaction to changes in any parameter depends on the strength of the coupling terms' response relative to the strength of the loss terms. Specifically, when a relevant biological parameter increases the flux of material through the coupling terms relative to the loss terms, the dynamics can be expected to become excited and destabilized. Similarly, as the inter-specific interaction strengths of the C-R interaction matrix (nondiagonals or a_{ij} terms) grow relative to the intraspecific interaction strengths (diagonals or a_{ii} terms), the dynamics tend to become more excited and less stable. This same result materializes in whole matrices and enables us to make sense of the results from modular and whole-community theory.

To see how this works, let us briefly visit the simple graphical techniques of Gershgorin discs employed by Haydon (1994, 2000).

10.3   GERSHGORIN DISCS FOR COMMUNITY MATRICES: AN INTUITIVE APPROACH TO EIGENVALUES

Let us consider an $n \times n$ real community matrix A, with entries a_{ij}. Here, n is the number of species, and the a_{ij} are Robert May's interaction strengths

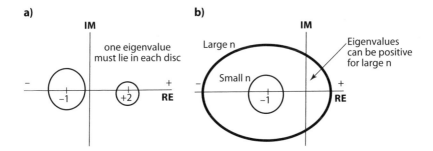

FIGURE 10.1. (a) Circles from the simple community matrix. The centers of the discs correspond to the diagonal, and the radius corresponds to the sum of the rows (columns could also be used). The two eigenvalues will neccessarily lie in the discs somewhere. (b) May's experiment pictured in Gershgorin discs. Notice that a small n gives smaller radii, and so the potential for having one large, positive eigenvalue is absent. Ultimately, increasing n increases the radius and so increases the likelihood of attaining unstable configurations (i.e., a larger positive eigenvalue).

as defined above and in chapter 5. Now let us use this community matrix to create the Gershgorin discs. Once created, these discs immediately give us an idea of the magnitude and sign of the actual eigenvalues that emerge from any dynamical system represented locally by a given matrix.

Gershgorin discs are plotted on the complex plane [e.g., a complex plane has a real value (the x-axis) and a complex value (the y-axis)]. For our purposes any point on this plane depicts a potential eigenvalue (recall that an eigenvalue is composed of a real value and a complex value). This graphical technique produces as many discs as there are species in our whole-community matrix, so n species gives us n discs. Each disc is created by centering the disc for a given row at its diagonal term (i.e., intraspecific interaction strength determines its place on the x-axis), and the radius of any disc equals the sum of all nondiagonal values in that same row (i.e., the sum of the inter-specific interaction terms yields the radius of the disc). Doing this n times for each row therefore creates n discs plotted on the complex plane for a given matrix.

As an example, take the simple 2×2 matrix

$$\begin{bmatrix} -1 & 2 \\ 0.5 & 2 \end{bmatrix}, \tag{10.2}$$

which corresponds to the two Gershgorin discs depicted in figure 10.1a. Gershgorin showed that every nonzero eigenvalue of the community matrix

A lies within at least one of the Gershgorin discs. In figure 10.1, then, we know that all discs must contain an eigenvalue. We do not know exactly where it is in that disc, but we know it must reside in each disc. Thus, we know from the discs in figure 10.1 that one eigenvalue is unstable (positive) and the other is stable (i.e., negative). This matrix is locally unstable given the Gershgorin discs it produces.

The above is a simple case of how these discs allow us some insight, but Haydon (1994, 2000) used these discs in a way that shed light on Robert May's famous result that increased diversity destabilizes food webs (May, 1974a). First, consistent with May, Haydon chose all intraspecific interaction strengths to be -1. This means that all Gershgorin discs are centered at -1 (figure 10.1b). Second, May randomly chose interaction strengths from a statistical universe with a given mean interaction strength. This means that as one increases diversity, the size of the Gershgorin discs ought to tend to increase (figure 10.1b). Remember that the size of the disc corresponds to the sum of the interspecific interaction strengths. Because these interaction strengths are randomly chosen (each nondiagonal has the same mean and statistical properties), the bigger matrices sum larger rows and so necessarily tend to produce bigger discs (figure 10.1b). Figure 10.1b shows an example of a low-diversity disc centered at -1 and a high-diversity disc centered at -1.

One can see that as the discs grow, this pushes the discs with large diversity further and further into the positive real part of the plane. Because each disc *must* contain an eigenvalue, these big discs can now contain an eigenvalue that is positive (i.e., has an unstable configuration). Given that the matrices were populated with random values, Haydon (1994, 2000) argued that *n*-species matrices sampled over and over tend to produce eigenvalues that fill these discs uniformly. Thus, the more diverse systems tend to produce more unstable configurations (a lower PSW).

This is our first glimpse that increasing diversity is destabilizing. If the distribution of eigenvalues in these matrices falls uniformly in these discs, we have shown that diversity ought to be destabilizing. Haydon (1994, 2000) finished this analysis by showing numerically that this indeed happens with enough simulations. If we increase the magnitude of the nondiagonal terms (i.e., increase the interspecific interaction strength) and leave the diagonal terms unchanged, one expects to see the same result. Strong interspecific interaction strengths tend to be destabilzing. In a sense, both the diversity experiment and the interaction strength experiment increase the strength of the nondiagonals relative to the diagonals' terms and thus tend to destabilize the Jacobian matrix (discussed in chapter 5). This result thankfully agrees with consumer-resource theory. Thus, as May (1974b) originally pointed out, all else equal, increases in diversity or in interaction strength tend to decrease food web stability.

10.4 A CONTROLLED APPROACH TO FOOD WEB MATRICES

Although Gershgorin discs are helpful in understanding Robert May's statistical universe experiments, they become trickier to interpret when the sampled matrices produce eigenvalues that do not necessarily uniformly cover the area of the Gershgorin discs. This nonuniform production of eigenvalues can occur when the matrices have additional biological properties placed on them (Haydon, 2000). Here, I follow up on some of the suggestions of Haydon (2000) to argue that there are some rather intriguing "gravitational-like" properties of Gershgorin discs for some important biologically motivated matrices. I will rely largely on an empirical assessment of these intriguing properties, as I am not going to mathematically explore why they happen. However, Gabriel Gellner and I are currently working on a method that more explicitly derives these gravitational-like properties using the characteristic polynomial equations that underlie community matrices (Gellner and McCann, forthcoming).

Let me highlight these ideas concretely with some examples by employing a generic (canonical) ecological matrix with the following biologically motivated assumptions:

$$\begin{bmatrix} D_1 & -b \\ bf & D_2 \end{bmatrix}, \tag{10.3}$$

where D_i are some number of diagonal interactions with the same intraspecific strength, $-b$ is the mean strength of the interaction of predator on prey, and bf is the mean strength of prey on predator. Thus, this canonical matrix represents a food web matrix with half the species containing diagonals D_1 and the other half of the species containing different diagonals D_2 (let us define the distance between these as $D = D_1 - D_2$). The upper nondiagonals are negative and constant with mean strength, b, while the lower diagonals are positive and are often some fraction of b (fb, where f is a scaling parameter). The f-value represents a scalar that indicates how much the negative impact of predator on prey translates into positive growth of the predator. Some argue that this is significantly less than 1, but in cases where there exists an inverted pyramid (e.g., $C^* > R^*$), the scaling parameter f can be greater than 1 for ecological models. This matrix therefore mimics a general food web of potentially many consumer-resource interactions.

Following Haydon (2000), I have structured the diagonals to contain two subwebs of different underlying stability properties. Specifically, one set of

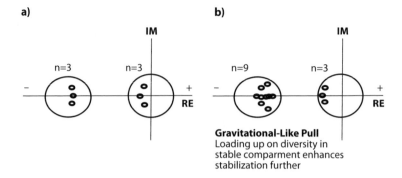

FIGURE 10.2. Discs from the canonical community matrix. (a) A six-species community with three strongly self-damped (discs with a more negative real part) species and three undamped species. Eigenvalues shown within discs tend to be negatively skewed for the less regulated subsystem and positively skewed for the strongly self-regulated subsystems. (b) Increasing the number of species in the strongly self-damped category acts to pull harder on the weakly regulated system such that it enhances stabilization further. This increased "gravitational" pull appears to consistently operate in these simulations.

species is going to tend to have more damping then the other, and so one set of species ought to be more inherently stable, all else equal.

If we take the matrsices as above and separate the diagonals by $D = -1$ (specifically, $D_1 = -1$ and $D_2 = 0$), we find that the Gershgorin discs centered on 0 tend to contain eigenvalues that are skewed to the left (figure 10.2a). Relative to the uniform case of Robert May, the eigenvalues are stabilized as though the more stable disc has a gravitational force that pulls at the eigenvalues in the zero-centered discs (figure 10.2a). This result is related to the fact that the sum of the real parts of the eigenvalues produced by the matrix, A, always equals the sum of the diagonal elements (trace) in A. Thus, if some eigenvalues are pulled left in the less stable discs, the eigenvalues must be pulled right in the stable discs to balance out the total effect (figure 10.2a).

The two underlying subsystems, therefore, have influenced each other such that the less stable subsystem is stabilized and the more stable subsystem is destabilized (Haydon, 2000)—there is a gravitational-like pull on the eigenvalues. Further, and still analogous to the gravitational metaphor, if we pack up a few more discs at $-D$ (i.e., we add a few more species to our stable subsystem or compartment in the whole matrix), we find that the eigenvalues in the zero-centered circles are pulled to even more negative values (figure 10.2b). The more dense the species are in the stable compartment, the more

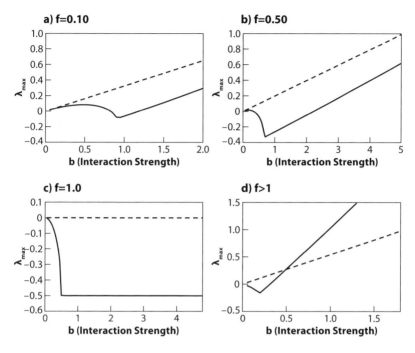

FIGURE 10.3. Response of a dominant eigenvalue (solid curve) as interspecific interaction strengths are increased for four values of f. (a) $f = 0.10$. (b) $f = 0.50$. (c) $f = 1.0$. (d) $f > 1.0$. The dotted curve represents the eigenvalue response of the unstable compartment by itself. Parameters: $n = 6$; $b = 0.10$; $D_1 = -1.0$; $D_2 = 0.0$.

this subsystem acts to tug at the unstable compartment and increasingly stabilize it. The above results are related to the weak interaction effect in the sense that a stable underlying subsystem can act to mute the unstable potential of the less stable subsystem.

To complete the analysis of this highly connected but compartmentalized ecosystem model, let us look at the influence of interaction strength, b, for different f values (the asymmetry of the consumer-resource interactions). Together these parameters comprise all the coupling terms in system (10.1). Figure 10.3a–d shows the results for different f values over a range of interaction strengths. Several things are clear. Coupling these two subsystems (solid curve) tends to stabilize the whole system relative to the most unstable subsystem (dashed curve). Weak to intermediate interaction strengths, show the most stabilization in terms of magnitude, while the greatest overall stabilization effect occurs when the off-diagonals are perfectly symmetric ($f = 1$; figure 10.3c). For the $f = 1$ case, it can be shown that the real eigenvalue is $-D/2$. The only case where the coupled system is less stable than the least stable subsystem occurs in cases where f is greater than 1 (figure 10.3d). Even here, though, weak interactions stabilize the system. Again, $f > 1$ effectively means that the ecosystem has inverted biomass pyramids.

While the above results were produced with 10-species simulations, they can also be produced with large matrices. In the $f = 1$ case it can be shown that this always produces a maximal stability curve like that shown in figure 10.3c. Further, the dependence on f remains qualitatively similar; there is no change in the nature of the stabilization result with small versus large matrices. Granted, as the diagonals become noisy and show little evidence of compartmentation, the results return to the historical finding that increased diversity is largely destabilizing. This points out that interaction strength structure can be critical to understanding stability. I will now revisit some classic whole-community results from the perspective adopted above.

10.5   SOME CLASSIC WHOLE-MATRIX RESULTS

10.5.1   FEEDBACKS AND THE STABILITY OF COMMUNITY MATRICES (LEVINS, 1968; DAMBACHER ET AL., 2003)

Here, I briefly review some of the work of Levins (1968) and the more recent work of Dambacher et al. (2003). Together their results aid us in understanding the connections between the stability of lower-dimensional modules and whole food webs. In a seminal contribution, Levins (1975) phrased the Hurwitz stability criteria in terms of the underlying feedbacks within a food web or community. In a greatly simplified sense, these feedbacks can be thought of as the dynamical feedbacks of the underlying modules that we have referred to in previous chapters.

Recall from previous chapters that a matrix is first turned into the characteristic polynomial, which coefficients $a_0, a_1, \ldots a_n$, from which one then determines the eigenvalues. Hurwitz's principal theorem, as restated by Dambacher et al. (2003), says that a matrix is stable if

(i) *All Polynomial Coefficients $(a_0, a_1, \ldots, a_n)$ in the Characteristic Polynomial Have the Same Sign*

A comment is in order. Levins (1975) and Dambacher et al. (2003) called each a_i in the characteristic polynomial the feedback F_i. By defining the arbitrary feedback a_0 as -1, the matrix is stable if all feedbacks are negative (because all must be the same sign and a_0 is negative). While negative feedbacks tend to enforce stability, there are ways this stability can be broken. The second Hurwitz criteria considers this aspect of destabilization (Dambacher et al., 2003) and can be stated as follows:

(ii) *The Hurwitz Determinants, $\delta_2, \ldots, \delta_{n-1}$, Are Positive*

Where the Hurwitz determinant is defined as in Dambacher et al. (2003). For the purposes of this book I do not want to focus on the precise definition of the determinants, because it is rather complicated, but instead give you a feel for its biological significance. To this end, we can follow the work of Dambacher et al. (2003) and state this second criterion less formally and more biologically. Effectively, criterion (ii) means that the lower-level feedbacks and upper-level feedbacks are properly balanced such that the higher-level feedbacks cannot be so large as to overwhelm the lower-level feedbacks. To begin to comprehend this second criterion, let us consider the three-species example of a food chain with omnivory.

The three-species omnivory matrix has three levels of feedbacks: (1) single-species intraspecific feedbacks ($F_1 = a_{ii}$ loop in a matrix); (2) two-species pairwise feedbacks (e.g., $F_2 = a_{ij} a_{ji}$ loops); and (3) three-link feedbacks (e.g., $F_3 = a_{12}a_{23}a_{31}$; note that this feedback occurs because of omnivorous pathways). Given that criterion (1) is maintained, we already know all the feedbacks, F_1, F_2, and F_3, are negative. It turns out that the second criterion in this three-species example says that the system is stable if and only if the feedbacks are constructed such that [see Levins (1975) or Dambacher et al. (2003) for more explicit discussions pertaining to the actual derivations of criterion (ii)]

$$\delta_2 = F_1 F_2 + F_3 > 0.$$

The $F_1 F_2$ is positive, so this means that the system can be made unstable only if F_3 is a big enough negative number such that it outweighs the positive $F_1 F_2$. Hence, a strong negative higher-level feedback can make this system unstable.

As a specific example of how this works, let us consider a Lotka-Volterra food chain model with resource damping. This model satisfies sign stability, and so it is always stable without omnivory ($F_3 > 0$). In this case, therefore, both Hurwitz criteria are satisfied. However, with the omnivorous link, the above criteria suddenly changes in that mathematically the system can now be unstable. It turns out this loss in stability is most likely to occur when the omnivorous links are strong (F_3 is big and negative), because this can make $F_1 F_2 + F_3 < 0$. Following this simple but inciteful observation, Dambacher et al. (2003) pointed out that three-level omnivorous feedbacks ought to be relatively weak if a system is to be stable. This is a hint that the matrix result may produce the same result as the omnivory module results discussed in previous chapters. This approach does not tell us how stable the matrix is, just whether it is stable or not.

As discussed in chapter 8, Gellner and McCann (forthcoming) developed a conceptually simple methodology, similar to the bifurcation techniques of modular theory, that enables one to consider how stability changes as one varies elements in a community matrix (i.e., one varies Robert May's inter-action strength metric). Gellner and McCann (forthcoming) showed that the general community matrix approach finds that weak interactions frequently, but not always, stabilize a food chain.

10.5.2   COMPARTMENTS (PIMM, 1979)

When May (1974a) found that diversity did not correlate with stability, he pos-tulated a number of biological mechanisms that might act in real systems to counter this seemingly contradictory result. One of his suggestions was that compartments in food webs—isolated subsystems with minimal linkage to other isolated subsystems—may stabilize diverse systems [also see Gardner and Ashby (1970)].

To explore this postulate, Pimm (1979) once again employed the classic matrix approach. He constructed a theoretical experiment in which he compared two completely isolated modules containing omnivory (i.e., com-pletely compartmentalized subsystems with same-chain omnivory) with a case where the omnivorous links instead occurred between the compartments. In this way he preserved the number of connections and made a completely compartmentalized web (i.e., the within-chain omnivory case) and a less compartmentalized food web (i.e., the across-chain omnivory case). The com-pletely compartmentalized cases were less stable in that they less frequently produced stable configurations. At a level this is not surprising, as webs with same-chain omnivory lose their sign stability property (i.e., they no longer sat-isfy the Hurwitz criteria as discussed above), while chains without omnivory are inherently sign-stable. Omnivory across chains preserves sign stability in these matrices.

Consistent with this theory, Pimm and Lawton (1980) found little evi-dence for compartmentation in early food web data. They were careful to point out that spatial habitat divisions can drive compartments and high-lighted an example from a spatially expansive arctic ecosystem. Similarly, Moore and Hunt (1988) made compelling arguments for compartments in soil food web data, an area we have discussed in previous chapters. Here, it is worthwhile noting that bacterial and fungal pathways occupy slightly different microhabitats in the soil, one more moist than the other (J. C. Moore, personal communication). Thus, both early empirical arguments for compartments had a common thread in that spatially distinct habitats may yield some degree of subweb differentiation or compartmentation. Moore and Hunt (1988) described compartments that happen at a fine scale within tiny habitat distinctions in the soil, while Pimm and Lawton (1980) discussed a spatially expansive example that occurs at the landscape scale. Further examination by others seems equivocal. Raffaelli and Hall (1992) found no strong evidence for compartmentation but cautioned that additional analysis was likely needed, while Solow and Beet (1998) found that webs were tightly connected within compartments and loosely connected between compartments relative to randomly constructed webs.

Recent empirical results have sought to more thoroughly explore the role of interaction strength in compartments. If fine-scale compartments exist, it may be that they are not topologically distinct but rather materialize only when we consider interaction strength. As an example, a food web researcher may find a highly connected whole web in nature, but this same web may be structured such that it is connected between subwebs largely by weak interactions. Krause et al. (2003) used data with estimates of interaction strength and found compelling evidence for food web compartments. Here, the interactions between subwebs were fewer and also tended to be weaker. The result is not really inconsistent with that of Pimm and Lawton (1980) because spatial divisions between compartments (e.g., pelagic versus benthic) seemed to also exist in the different compartments.

Some recent theory has found that relatively weak to intermediate connections between compartments can be stabilizing (Teng and McCann, 2004), especially if compartments are spatial and coupled by predators capable of responding rapidly to variability in the underlying compartments (McCann et al., 2005). This latter theoretical result, based on a spatial food web model, therefore suggests that we expect compartments between very distinct habitats ought to exist. Following this line of reasoning, it may also be that compartmentation is less likely in spatially constrained habitats where predators/consumers are capable of strongly coupling mutliple habitats and blurring habitat boundaries. So far as I know, this is an unexplored empirical question.

To further examine compartmentation for more complex food webs, I created random community matrices with 10 species following the canonical matrix formulation outlined above in system (10.1). I then varied the strength of all interactions that cross between the strongly regulated subsystem (the 5 more negatively constructed diagonals) and the less regulated 5-species subsystem. The D_i values again identify the underlying compartments.

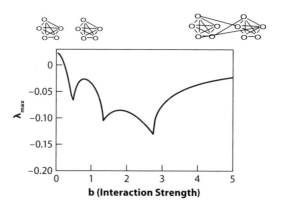

FIGURE 10.4. Maximum eigenvalue response of a 10-species web connected as discussed in the text. Throughout the experiment all links are maintained, however, the strength of between-compartment links are varied from 0 to 5. Generally speaking, the food web is first stabilized and then destabilized, as in previous results. However, with greater diversity come more peaks and valleys in the stability response curve. Parameters: $b = 0.30; f = 0.80; D_1 = -2.0; D_2 = 0.0$.

Figure 10.4 shows the eigenvalue results: For intermediate to weak couplings between compartments, the whole system is stabilized; however, as the cross-subsystem coupling gets strong, the whole system is ultimately destabilized. Here, the empirical patterns of the Gershgorin discs discussed above operate for weak to intermediate couplings. That is, the stable compartment pulls the eigenvalues of the less stable compartment toward the negative side of the Gershgorin discs (see figure 10.2a and b). At great enough interaction strengths, the Gershgorin discs swallow each other and the gravitational-like properties of the once-stable subsystem no longer operate.

The stability curve in figure 10.4 shows numerous peaks and valleys, although the general pattern of stabilization followed by destabilization remains. I have no real idea why this happens but speculate that it may be related to the creation and destruction of complex signatures in the more reticulate web.

There remains the question of how the level of connectance itself influences stability. In the above experiment I kept the system completely connected and varied only cross-compartment interaction strengths. What if we instead varied the number of connections between compartments and asked how the connectance between compartments influences stability? To do this I increasingly zeroed out interactions between compartments but kept mean interaction strengths between compartments constant. It seems here too that intermediate amounts of coupling might offer the most stabilization (figure 10.5). Theory

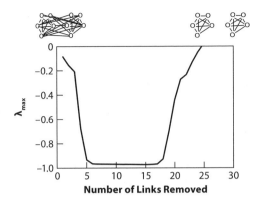

FIGURE 10.5. Maximum eigenvalue response of a 10-species web connected as discussed in the text. Initially compartments are completely coupled such that all links exist. I then look at the impact of removing 1 to 25 links on stability. Links are randomly removed, and the average eigenvalue response is plotted as a function of the number of links removed. Parameters: $b = 0.30$; $f = 0.80$; $D_1 = -2.0$; $D_2 = 0.0$.

that has not found compartments stabilizing [e.g., Pimm and Lawton (1978); Solow et al. (1999)] has also examined completely compartmented webs (i.e., there was absolutely no connection between webs) versus random webs. The results above agree that compartments are not stabilizing when they become completely isolated but are stabilizing for modest to intermediate numbers of connections.

10.5.3   STABILITY, ASSIMILATION EFFICENCY, SELF-REGULATION, AND DONOR CONTROL (DEANGELIS, 1980)

DeAngelis (1975) reviewed the results of Gardner and Ashby (1970) and May (1974b) to argue that diversity can be stabilizing under certain conditions. Specifically, he showed that if food webs are biased toward low assimilation efficiencies, strong self-regulation, and donor control, stability can actually increase with diversity. Such biological assumptions make these system less excitable (i.e., eigenvalues become more dominated by monotonic solutions), and so increased diversity can stabilize complex webs. Another way of stating this is that all the mechanisms proposed by DeAngelis (1975) weaken the influence of the nondiagonal coupling terms relative to that of the diagonal terms in the community matrix. As discussed from chapter 5 on, this weakening of the nondiagonal terms is a potent stabilizing force. This result resonates with the general idea that muted interactions can reduce the likelihood of runaway growth and overshoot dynamics, promoting the stability of even complex food webs.

10.6   RECENT WHOLE COMMUNITY APPROACHES

10.6.1   Neutels Weak Loops (Neutel et al., 2002)

Recent work on the stability of soil food webs has found that empirical data suggest that long feedback loops often contain weak interactions. Using a novel metric of stability, Neutel et al. (2002) argued that these weak loops were stabilizing relative to similar webs where the same set of interaction strengths were randomly assigned. In a sense this result speaks to both the weak interaction effect and the feedbacks of Levins (1975) and Dambacher et al. (2003). Hurwitz criteria (ii) states that higher-level feedbacks cannot outweigh lower-level feedbacks. The result of Neutel et al. (2002), therefore, ensures that the weak interactions in long omnivorous loops weaken higher-level feedbacks and so tend to produce matrices that satisfy Hurwitz criteria (ii).

10.6.2   Strong Interactions are not Destabilizing (Allesina and Pascual, 2008)

Allesina and Pascual (2008) suggested a result that appears counter to many results discussed in this book. Specifically, they found that strong consumer-resource interactions do not tend to be destabilizing. Further, because of this result, they argued that topology, not interaction strength, largely governs the stability of community matrices. At first glance this result is mystifying, as it counters May's classic work as well as an enormous body of research (including C-R theory) that has effectively recreated May's classic result over and over again [e.g., Haydon (1994, 2000)]. It turns out that at least one significant difference between the Allesina and Pascual (2008) result and historical work is that Allesina and Pascual (2008) manipulated both interspecific interaction strengths (the off-diagonals) and intraspecific interactions simultaneously (the diagonals) when they increased interaction strength.

   In their theoretical experiment, they increased the absolute value of the negative diagonal or damping terms as the nondiagonal interaspecific terms were increased. If we recall all the stability results discussed to this point, the stability of such a matrix tends to be decreased if the off-diagonal terms are strengthened. Additionally, the stability of food webs tends to increase if the intraspecific damping terms (diagonals) are made more negative. Thus, if we increase the absolute magnitude of both interspecific and intraspecific interaction strengths simultaneously, we do not necessarily expect to see the destabilization result of May or others (recall that they tended to hold the

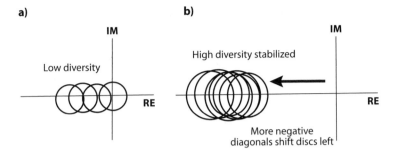

FIGURE 10.6. Graphical depiction of increasing strength of intraspecific (nondiagonal) terms in the community matrix while simultaneously making the intraspecific terms (diagonals) more negative. (a) Low-diversity system. (b) High-diversity system. Here, the negative terms push the Gerschgorin discs to more negative, more stable values. This shift outweighs the destabilizing increase in disc size that accompanies more diversity, as discussed previously.

diagonal terms constant). I argue that this is what happens in the Allesina and Pascual (2008) result—the growth of the damping terms outweighs the destabilization of the interspecific coupling terms.

To see this let us again consider Haydon's use of Gershgorin discs. The discs in this case grow (increasing interaction strength, b, inflates the disc size), and thus the ability for any eigenvalue to become more positive increases. Now if we also simultaneously increase the negativity of the diagonals, we act to shift the Gershgorin discs left at the same time (figure 10.6a and b). Clearly, if the negative shifts outweigh or equal the growth of the discs, we expect that the results that increase in interaction strength are not destabilizing. It is therefore true that interaction strengths do not increase stability if, on average, they are countered by increased self-regulation. This result is therefore completely consistent with the food web theory presented throughout this book, however, it represents a specific set of biological assumptions (i.e., the diagonals damp with greater intensity when interspecific interaction strengths, b, increase).

Given this result, it remains empirically critical to ask if such increased damping tends to accompany increased predator-prey interaction strengths. I do not know of any empirical result that would suggest this, but it remains a possibility. This result, in hindsight, is consistent with DeAngelis' assertion that strongly self-regulated communities can be stabilized by diversity.

10.6.3   THE STABILITY OF PLAUSIBLE FOOD WEBS

A number of researchers have sought to understand whole-community dynamics by constraining the parameter set that underlies models (Yodzis, 1981; Yodzis and Innes, 1992) . This has often been done by making body size arguments. Yodzis (1981), for example, pointed out that plausible inter-action strength estimates from real webs produced more stable configurations than those from random webs. Other results have followed. de Ruiter et al. (1995) examined a range of soil webs and found again that the patterning of the empirically motivated interaction strengths tended to produce stable dynam-ics. Altogether, these results suggest that the natural patterning of predator and prey in real ecosystems affects the arrangement of interaction strengths, which in turn determines food web stability. These more historical results hint that body size may be a fundamental axis governing this result.

Similarly, Gross et al. (2009) set out to create a biologically motivated class of matrix models that speak both to the classic whole-system matrix approach and to more recent modular food web approaches. Gross et al. (2009) did this by constraining the statistical universe of interaction strengths via assump-tions about underlying growth, loss, and consumption processes. By doing so, they in essence constructed a matrix approach that ought to be more consis-tent with modular theory because their biologically motivated matrices have more in common with specific food web model formulations. Nonetheless, their approach still employed the classic type of question. That is, they con-sidered the answer from randomly generated parameters and monitored the percentage of stable webs. As a result, they are not comparing how certain structures mediate stability relative to some control.

Their results confirm much of food web theory, but not all. The authors, for example, considered the effect of looking at the influence of variation in interaction strength on the dynamical outcomes. Here, they looked at the influence of the coefficient of variation (variance/mean) of interaction strength by holding the mean interaction strength constant and changing the variation around the mean interaction strength. By doing so, they are able to identify the role of just variance in interaction strength on stability. In this case, Gross et al. (2009) found that high variation in interaction strength in small matrices tended to correlate with increased stability, while in larger matrices the answer was reversed with increasing variance in interaction strength, reducing the likelihood of finding stable webs. This may be seen as an argu-ment against weak interactions playing a role in large systems, and it might be. However, this approach—a clear attempt to unite theory—still potentially misses out on important underlying contingencies. As an example, despite the finding that unstable configurations occur more frequently in large webs than in small webs, it still may be true that weak interactions play a power-ful stabilizing role in diverse networks. The diverse webs, though, may just have a larger destabilization phase than stabilization phase with increases in interaction strength. Further work resolving this result mechanistically would be beneficial.

Finally, much research has considered the implications of the growing body of network theory on food webs (Dunne et al., 2002). Here, researchers have had success using simple models to reproduce food web topologies found in real webs [e.g., the niche model; Williams (2000)]. Using plausible estimates of coupling strength (Yodzis and Innes, 1992), they have also begun to investigate the dynamics produced by these simple but effective food web topology models (Martinez et al., 2006; Williams, 2008). These papers almost universally appear to find that predator-prey body size ratios are critically important for stability (McCann and Yodzis, 1994; Emmerson et al., 2004; Loeuille and Loreau, 2005; Woodward et al., 2005; Martinez et al., 2006; Brose et al., 2006a). Indeed, Brose et al. (2006b) showed that predator-prey body size ratios drive a positive diversity-stability relationship for empirically derived body size ratios.

In a related contribution, Otto et al. (2007) found that stability also may be driven by the topological properties imparted by body size. Otto et al. (2007) found that allometric degree distributions, in which large organisms tended to have more prey but fewer predators, played a fundamental role in stabilizing data taken from five natural webs. This latter result is also consistent with the recent work of Gross et al. (2009) who found that food web stability in complex systems was enhanced when (i) high trophic level organisms ate multiple prey, and (ii) species at intermediate trophic levels were fed on by multiple predators. All of the above results agree with the notion presented in previous chapters that larger higher trophic level organisms may play fundamental roles in mediating the persistence of complex webs. Further, the results agree with the historical suggestion that body size is a major role in determining food web stability and dynamics.

Finally, in a recent experimental test of weak interaction theory in a large marine field experiment, O'Gorman et al. (2008) found that removing either strong or weak interactors from an existing web weakened the stability of the ecosystem, resulting in cascading losses of some species. Their results are at least consistent with the idea that the patterning of interaction strengths plays a potent role in maintaining the diversity and function of ecosystems. This is one of the larger-scaled experiments on food web interaction strengths since the pioneering work of Robert Paine, who found that a top predator, the starfish *Pisaster pisaster*, played a major role in maintaining the diversity of coastal benthic food webs via differential predation on competitors (Paine, 1992). In a sense, the result of Paine (1992) was an empirical example of a complex diamond-like module with differential interaction strength pathways.

10.7   SUMMARY

1. All in all, theory from many complex species systems tends to resonate with ideas from C-R theory and modular theory. As an example, weak interaction stabilization results recur repeatedly in both simple and more complex food web models. Some differences, nonetheless, clearly exist. This may be due in part to the slightly different questions being asked. For example, community matrix approaches often consider the percentage of stable webs, whereas modular theory considers how aspects like interaction strength change stability relative to some control case.

2. Scaling from small modules to whole systems may be accomplished by understanding how adding lower-level modules influences the flow of energy through potentially unstable modules. Here, similar to Haydon (2000), I showed that stable systems can often stabilize unstable systems. Similarly, weakly coupling compartments of different underlying stability acted to stabilize the dynamics of the least stable subsystem.

3. Historical and recent evidence exists [e.g., Pimm and Lawton (1980); Krause et al. (2003)] that distinct habitats may tend to produce compartments. This appears not as clearly evident at the microhabitat scale as it is in the larger-scaled spatial divisions (Montoya et al., 2006).

4. Future research extending or comparing modules to complex whole systems may be able to utilize some of the interesting ideas on subsystem feedbacks in complex systems utilized by Levins (1975) and Dambacher et al. (2003). This work, for example, points out that the higher-order feedbacks (long loops in food webs) must be weak to satisfy the stability criteria of Hurwitz. This result resonates with the more empirically motivated results of Neutel et al. (2002) that find weak links in long loops in soil food webs.

5. Researchers have increasingly sought to constrain parameter spaces when exploring whole food web models. Recent advances using body size relationships have found that nature often occupies a "parameter space" that is stable relative to random food web constructions. Further, higher-order predators that are generalists have been frequently shown to play a critical role in mediating food web persistence.

CHAPTER ELEVEN

Adding the Ecosystem

To this point in the book, I have completely ignored the role of nutrient recycling and decomposition. Nutrient recycling and decomposition form larger-scale feedbacks at the ecosystem scale that may play a significant role in the dynamics and stability of food webs. In this brief chapter, I start to explore this fundamental problem and reconcile previous chapters with some of the pioneering efforts of O'Neill (1976) and DeAngelis (1980, 1992) that appear to counter some of the ideas in food web theory. In what follows, we review some of the existing theory on detritus and food web dynamics. I will show that the results, as stated, suggest an inconsistency between these two theories. I then revisit detrital models to show that these apparent differences are reconcilable. Finally, I will argue that detritus is largely a stabilizing force in food webs. Detritus operates in both equilibrium and nonequilibrium dynamics by distributing the nutrients to the classic food web in a manner that tends to weaken top-down control in ecosystems relative to that in community-based models.

11.1 GRAZING FOOD WEBS AND WHOLE ECOSYSTEMS

Classic food web approaches have ignored the role of detritus in dynamics (Moore et al., 2004). However, ecologists interested in the role ecosystems play in governing dynamics have made some clear theoretical predictions (Jordan et al., 1972; O'Neill, 1976). Specifically, DeAngelis (1980) argued that the resilience of an ecosystem (as measured by the size of the eigenvalue) tends to be governed by the magnitude of the flux per unit of standing crop and the *index of recycling* (defined as the average number of times a unit of material is recycled before leaving the system). DeAngelis (1980) further postulated that these two measures can be encompassed by a single metric, the *nutrient turnover time* (defined as the total nutrients, N_T, in the system divided by the influx of nutrients, I_N). A generalization of his results can be stated as

(i) "The study of energetic models has led to the conclusion that the magnitude of the flux, in this case energy or biomass, per unit standing crop in the system, or the power capacity, is positively correlated with the system's resilience" (DeAngelis, 1980), and

(ii) "the mean transit time should, in principle, be strongly positively correlated with the recovery time of an ecosystem model from a perturbation and therefore inversely correlated with the resilience" (DeAngelis, 1980).

The above results suggest that increased flux rates, either between species or entering the nutrient pool, tend to stabilize the system. This is in stark contrast to many of the ideas that we have entertained so far. That is, weak flux rates between interactions or reduced production often tends to stabilize. Recall though, that the population and food web models of the previous chapters gave the same result when the dynamics were not excitable (i.e., they were monotonic, with real eigenvalues and trajectories that did not fluctuate on their way to the equilibrium). In most of these models, the space of monotonic trajectories was limited relative to the excitable portion in parameter space.

In what follows, I will argue that DeAngelis' result is an interesting case where under specific biological conditions (an ecosystem is nearly closed) the dynamics tend to lose these complex oscillatory signatures (i.e., the eigenvalues tend to be real in the DeAngelis case). As a result, one expects that increased flux rates ought to be stabilizing in closed systems. Consistent with this, Nakajima and DeAngelis (1989) found that a nutrient-based model with saturating uptake and consumption rates, as well as tight recycling, eventually was destabilized by nutrient enrichment, I_N, but that this onset of destabilization occurred just prior to the Hopf bifurcation. These results suggest that the nearly closed ecosystem delayed the destabilization phase significantly.

Nonetheless, oscillations or oscillatory decays exist in many natural systems [Kendall et al. (1998); see chapters 4 and 5], and so the above result demands a little further theoretical consideration because it seems that this prediction of the closed ecosystem model does not always hold. Further, and seemingly at odds with this result, DeAngelis (1980) found that making an open ecosystem more closed tended to be destabilizing (Loreau, 1994).

To look into this more I chose to start with two models of the structure identified in figure 11.1. Figure 11.1a shows a classic food chain representation in which nutrients are taken up by a resource that is in turn eaten by a consumer. There is no explicit recycling of nutrients; there is merely an influx, I_N, and a

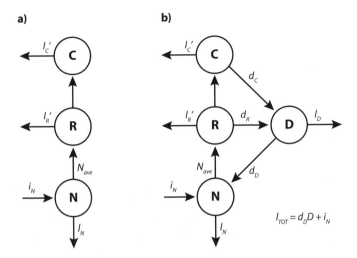

FIGURE 11.1. Schematic of the classic food chain model view (a) and the ecosystem variant, including recycling and decomposition (b), examined in this chapter.

loss of nutrients (a loss term exists in each component of the web). While this starts with the currency of nutrients, it is still similar in spirit to classic food chain approaches in that it does not recycle nutrients through the system. The model can be generally written as

$$
\begin{cases}
\dfrac{dN}{dt} = I_N - l_N N - rNR, \\[2mm]
\dfrac{dR}{dt} = rNR - (m_R + l_R)R - a_{CR}RRC, \\[2mm]
\dfrac{dC}{dt} = ea_{CR}RC - (m_C + l_C)C,
\end{cases}
\tag{11.1}
$$

where N, D, and R are the nutrient pool, detritus, and resource, respectively. The parameters are as follows: I_N is the nutrient input rate, d_D is the rate of decomposition, l_N is the loss due to nutrient leaking, r is the nutrient uptake rate of the resource, m_i is the nutrient loss of species i that is potentially recycled (see next model), l_i is the loss of nutrients from the pool, e is the conversion of resource nutrients into consumer nutrients, and a_R is the uptake or consumption rate.

Figure 11.1b, on the other hand, depicts the basics behind an ecosystem model that also contains the classic grazing interaction, C-R. Here we have a fraction of the loss rates moving through the detritus compartment and back to the nutrient pool at a rate that governs decomposition, d_D, yielding

$$
\begin{cases}
\dfrac{dN}{dt} = I_N + d_D D - l_N N - rNR, \\[2mm]
\dfrac{dD}{dt} = (1 - e)a_R CR + m_R R + m_C C - d_D D - l_d D, \\[2mm]
\dfrac{dR}{dt} = rNR - (m_R + l_R)R - a_{CR}RC, \\[2mm]
\dfrac{dC}{dt} = ea_{CR}RC - (m_C + l_C)C,
\end{cases}
\tag{11.2}
$$

where N, D, R, and C are the nutrient pool, detritus, the resource, and the consumer, respectively. The parameters are as follows: I_N is the nutrient input rate, d_D is the rate of decomposition, l_N is the rate of nutrient leaking from the system, r is the nutrient uptake rate of the resource, m_i is the nutrient loss rate of species i that is decomposed, l_i is the nutrient loss rate from the system, e is the conversion of resource nutrient into consumer, and a_{CR} is the uptake or attack rate on the resource by the consumer.

In order to set up the problem it is important to control for the influence of productivity. The detrital model has an additional parameter, namely, the detrital recycling rate, $d_D D$, and this link completes the cycle of nutrients [equation (11.2)]. This link therefore modifies the total amount of nutrients the N pool receives and the amount of productivity of R relative to the classic food web approach [equation (11.1)]. Specifically, given the same parameter values, the flux into the N pool in the detrital module is greater than that in the food web module (because $I_N + d_D D > I_N$). This aspect of the detrital model—that it increases the flow of nutrients back into the classic chain—is evident and not really what we seek to explore. We want to know how detritus influences C-R dynamics relative to the non-detrital model, all else equal. Further, and importantly, changing the productivity has a profound influence on the dynamics, and if we do not account for this, the remainder of my results may be simply due to the enhanced production the detrital module receives. In what follows, I will transform the N-R-C model parameter, I_N, such that the nutrient input term is always equal to the detrital influx rate plus the nutrient input (i.e., $I_N' = I_N + d_D D^*$, where D^* refers to the detrital density at equilibrium). I now briefly explore the N-R-D submodule before considering the full N-C-R-D model.

11.2   THE N-R-D MODULE

Let us first start by assuming that the consumer in system (11.2) is absent and explore the underlying N-R-D submodule. The closed system has a nice simple

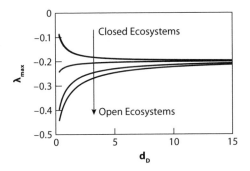

FIGURE 11.2. Maximum eigenvalue is followed as the detrital decomposition rate is increased for closed to more open ecosystems. Closed systems are stabilized by the increased recycling as per DeAngelis (1980), while more open systems are destabilized by the increased recycling. Parameters: $I_N = 4.0$; $r = 1.20$; $a_R = 4.10$; $m_R = 0.40$; $l_R = 0.20$; $e = 0.80$. Most closed to open systems are (i) $l_N = 0.05$, $l_D = 0.001$; (ii) $l_N = 0.20$, $l_D = 0.01$; (iii) $l_N = 0.50$, $l_D = 0.30$; (iv) $l_N = 0.80$, $l_D = 0.75$; (v) $l_N = 1.0$, $l_D = 1.10$.

result. It drives dynamics with little or no overshoot (i.e., the eigenvalues tend to be real). As such, the system is not excited by increased production through its pathways. From what we have learned to this point, one would expect this system to be powerfully stabilized by such increased fluxes. Figure 11.2 shows that for almost closed systems (i.e., l_N is small) the system is indeed stabilized by increased recycling rates. This agrees with the results of DeAngelis (1980, 1992).

However, if we relax the closed ecosystem condition, we quickly find that this is not always true. In fact, the more open system switches to becoming destabilized by this increased flux rate (the lower curves in figure 11.2). Note also that the open systems are inherently more stable than the closed systems, as DeAngelis (1980) and Loreau (1994) pointed out.

It remains to see if this destabilization is consistent with the theoretical ideas developed in previous chapters. There, we tended to find that complex eigenvalues yielded excitable dynamics—that is, dynamics that become more oscillatory with increased fluxes through any given species or interaction. In this case, though, the answer is modestly different. The destabilizations in the leaky ecosystem all show some evidence of complex subdominant processes that may belie the destabilizing force. While subdominant eigenvalues are not generally used for estimating stability, they can contribute to return times, $1/\lambda_{max}$ when the eigenvalues are of similiar order. However, the destabilizing region does not exactly map to the complex subdominant region in all cases, so some more things are going on here that drive this result. Nonetheless, the result that open systems become more excitable suggests that addition of the C-R interaction on top of this ought to only enhance this aspect in open versus closed detrital systems. I now consider the influence of this detrital module on a C-R interaction.

11.3   DETRITUS AND C-R INTERACTIONS

I now perform the same experiment as above but for the N-R-C-D module and compare the results to the classic community model without detritus (see figure 11.1). Further, I will consider this for both a relatively closed ecosystem and a more open ecosystem. To compare these different models, we ran the N-R-C-D model by varying the decomposition rate, d_D, and kept track of the total nutrient input entering the nutrient pool. This is simply the summation of all the fluxes into the N pool (figure 11b; i.e., $I_N' = I_N + d_D D^*$). In order to compare these results to the community model (N-R-C), I then ran the N-R-C model over the corresponding range in I_N. All parameters in the models are the same; only the ecosystem model gets some of its nutrient input via recycling, and the community does not have this feedback. This puts us in a position to ask what the role of recycling, in and of itself, is in stability.

When we do this, the first result is that the closed ecosystem actually can be less stable than the N-R-C model for low recyling rates and low total nutrient inputs, but generally the ecosystem model ends up more stable for high recycling rates and higher total nutrient inputs (figure 11.3a). Recall that the closed case is the DeAngelis case and suggests that increased flux rates ought to frequently stabilize. Here, the increased flux rates stabilize the detrital model, and as long as recyling is high enough, it pushes the N-R-C-D model to be more stable than the classic model (figure 11.3a). The detrital model eventually takes advantage of the high nutrients to stabilize the system.

For open systems, though, the detritus model in this case appears to be always more stable then the N-R-D model (figure 11.3b). This result suggests that detritus may be most poignant in stabilizing dynamics when the ecosystem is open. Somehow, and most curiously, the detritus is able to harness the potential instability (recall that the potential for complex eigenvalues or oscillatory decays in the N-R-D model increases with openess) and turn it into enhanced stability. I will return to explore this aspect of the stabilization later.

Attack rates in the C-R interaction mediate the strength of the oscillatory decay (chapter 5), and so we now ask how changing interaction strength through consumption rates influences these two different models. Again, the

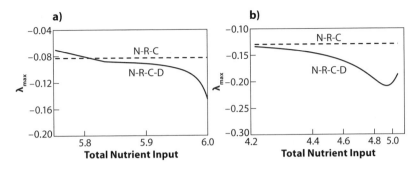

FIGURE 11.3. Stability of the community module (N-R-C; dashed curve) and ecosystem module (N-R-C-D; solid curve) are looked at for a closed (a) and an open ecosystem (b). (a) Closed parameters are $I_N = 4.0$; $l_N = 0.20$; $r = 1.50$; $a_R = 4.10$; $m_R = 0.30$; $l_R = 0.20$; $e = 0.80$; $m_C = 0.30$; $l_C = 0.40$; $l_D = 0.01$. (b) Open parameters are $I_N = 4.0$; $l_N = 0.50$; $r = 1.50$; $a_R = 4.10$; $m_R = 0.30$; $l_R = 0.20$; $e = 0.80$; $m_C = 0.30$; $l_C = 0.40$; $l_D = 0.30$. d_D varies in N-R-C-D to create total nutrient range for the N-R-C-D model ($I_N + d_D D^*$), and then this range of total nutrients is used as I_N for the N-R-C model.

results depend on the closed versus the open assumption: the N-R-C model being more stable for low to moderate attack rates on average, and the N-R-C-D model eventually being more stable for high attack rates (figure 11.4a). The N-R-C model experiences a discontinuous shift to a higher destabilization rate at an attack rate of approximately 1. This occurs when the complex subdominant eigenvalue shifts to being the dominant complex eigenvalue. This complex portion is likely contributed by the C-R interaction, and so it appears that the community model, although more stable for a while, is ultimately swamped by the destabilizing potential of the underlying oscillatory C-R subsystems. The N-R-C-D model is not shifted in such a manner in figure 11.4a, although it can be at high attack rates. This is generally true and suggests that the N-R-C-D model again somehow fights off the potential instability of an excited C-R interaction better than the nondetrital model. The more open detrital model rapidly becomes more stable than the community model (figure 11.4b). Again, all the above results hold the total nutrient flux into the nutrient pool constant, so none of them are due to productivity.

These ideas are consistent with some of the literature on this topic, even for nearly closed ecosystems. Specifically, Nakajima and DeAngelis (1989) used the same model above with the addition of a saturating uptake and consumption rates (I will consider this shortly). Nakajima and DeAngelis (1989) found that increasing nutrient input may at first lead to increased resilience (the DeAngelis result), but resilience decreases sharply immediately before

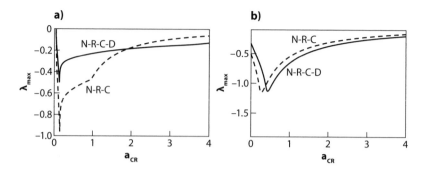

FIGURE 11.4. Stability of the community module (N-R-C, dashed curve) and ecosystem module (N-R-C-D, solid curve) are looked at for a closed and open ecosystem as the attack rate, a_C, on the resource is increased. (a) Closed parameters: $I_N = 4.0$; $l_N = 0.05$; $r = 1.50$; $a_R = 4.10$; $m_R = 0.30$; $l_R = 0.20$; $e = 0.80$; $m_C = 0.30$; $l_C = 0.40$; $l_D = 0.001$. (b) Open parameters: $I_N = 4.0$; $l_N = 1.0$; $r = 1.50$; $a_R = 4.10$; $m_R = 0.30$; $l_R = 0.20$; $e = 0.80$; $m_C = 0.30$; $l_C = 0.40$; $l_D = 1.10$. d_D varies in N-R-C-D to create the total nutrient range for the N-R-C-D model ($I_N + d_D D^*$), and then this range of total nutrients is used as I_N for the N-R-C model.

the Hopf bifurcation occurs. In other words, once the C-R interaction is destabilized, it overwhelms the closed system dynamics abruptly. As we saw above, the more open system appears to combat this onset of C-R instability a little better. We now consider why this can happen mechanistically.

This enhanced stabilization can come from only a few things. The detrital model has all the same rates, but with the addition of detritus, the equilibrium densities of the N-R-C components can change. As d_D increases to high rates, we expect the D pool, for example, to approach very small numbers reflecting the fact that the nutrients are tied up in the biotic biomass. At the other extreme, when d_D is near zero, the nutrients become tied up in the detrital biomass and the system survives solely on the new influx of nutrients, I_N. This rearrangement of biomass may play a role in the stabilization. A second possibility is that the detritus acts to liberate nutrients back into the nutrient pool in a manner that is not necessarily the same as in the classic model. The classic model effectively assumes a constant influx, I'_N, irrespective of the densities of its pool nutrients. Thus, after a perturbation the amount of nutrients entering the nutrient pool is constant. With detritus included, though, the actual amounts of nutrients entering the nutrient pool clearly become a function of the density of all its biotic members. Exactly how this pans out, though, is not immediately obvious. This latter aspect of detritus—its role in distributing nutrients after a perturbation—may be a particularly stabilizing force if

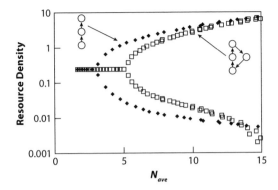

FIGURE 11.5. Bifurcation of the resource density across a range of average nutrient inputs. Communities containing a detrital module (open boxes) bifurcate at higher N_{ave} than communities without a detrital module (solid boxes). The amplitude of resource cycles expands further, and densities reach closer to zero, in communities with a detrital module.

it manifests as a negative density-dependent function. As an example, a car's distributor takes high voltage and distributes it temporally in an out-of-phase manner that makes the spark plugs fire serially. This consistent firing, perfectly out-of-phase, maintains the stable function of a car engine. We will see that this "distributor effect" may operate naturally in ecosystems.

11.4 NONEQUILIBRIUM DYNAMICS AND DETRITUS AS A DISTRIBUTOR

Recently, Jason Rip and I examined exactly this aspect (i.e., the distributor idea above) of detritus. To examine how detritus may act to distribute nutrients, we considered a model that begets nonequilibrium dynamics [i.e., we included type II functional responses in system (11.1)]. With the nonequilibrium dynamics we could easily ask whether the nutrients are distributed in a manner that excites oscillations further or mutes them.

The first result was that the detrital model tended to transition into instability later than the nondetrital model (figure 11.5). Parameters were chosen randomly 100 times, and this happened in all 100 cases. These results are consistent with the equilibrium ideas presented above. To assess how the detritus distributed the nutrients back to the N pool we next considered the correlation between N and R densities. Their relative phase relationship ought to have enormous impacts on dynamics because synchronized N and R dynamics ought to produce conditions of runaway growth when both N and R are high,

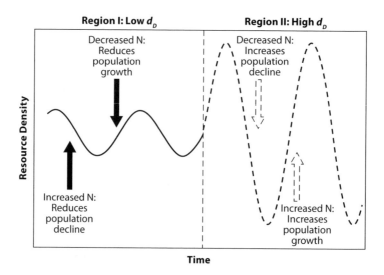

FIGURE 11.6. Effect of the detrital recycling rate on community stability (measured as the amplitude of the consumer cycle).

and heightened decay when they are both at low densities relative to the classic case where N is effectively decoupled from the R dynamics. Similarly, if detritus liberates nutrients, N, such that they are out of phase with R dynamics, we would expect a strong stabilizing potential (see figure 11.6 for a diagram of this phenomenom). Because we have found the tendency for stabilization, we expect this later out-of-phase mechanism to occur.

We checked the correlation as a function of the decomposition rates, d_D, and found that the maximum eigenvalue curve correlated strongly with the N-R correlations. Specifically, figure 11.7 shows that as stability increases, so does the degree of negative correlation. Further, we also found that the correlation is always more negative than in the classic nondetrital models.

The out-of-phase dynamics between the detrital and classic grazing subsystems suggest that a consumer capable of coupling into either channel may receive the benefit of this out-of-phase production (i.e., when resources are low, detritus production is high, so the consumer is buffered, and vice versa). While I have not examined this apsect here, the exhaustive bifurcation analysis by Edwards (2001) of a similarly constructed N-P-Z-D model (nutrient-phytoplankton-zooplankton-detritus) suggest that consumers (zooplankton) capable of coupling into both the detrital channel and the grazing channel are stabilized relative to the case where they do not couple into both channels. This idea agrees with previous chapters where I discussed how couplers capable of dipping into different compartments can be

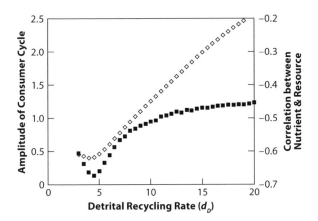

FIGURE 11.7. Effect of the detrital recycling rate on community stability (measured as the amplitude of the consumer cycle) and the correlation between nutrients and resources. There is a unimodal relationship between the recycling rate and stability, as well as between the recycling rate and the correlation between nutrients and resources.

stabilized, and this stabilization is likely enhanced when these compartments are asynchronous.

11.5   DISCUSSION

This chapter is a preliminary and cursory look at the influence of nutrients and detritus placed within the food web context laid out in this book. There is much that needs to be done, and the arguments here are presented more to place the work of DeAngelis (1992) and stimulate further research on bridging ecosystem to food webs. Often research groups seem to operate independently of each other despite working on areas of research that overlap greatly.

There is little empirical work that links ecosystem level theory to stability. The above results would benefit from an operational knowledge of where ecosystems are in terms of the detritus recycling rate and where natural ecosystems lie on the open-closed ecosystem continuum [e.g., Essington and Carpenter (2000)].

11.6   SUMMARY

1. Closed ecosystems tend to produce montonic, or unexcited, dynamics and so are generally stabilized by increased production as per DeAngelis (1980,

1992). Open systems are more excitable, show overshoot potential, and so can be destabilized by production.

2. The addition of a C-R incteraction to a simple detrital module (nutrients, resource, and detrital pools), even in a closed system, eventually can drive overshoot dynamics and destabilization by increased production, interaction, or coupling strength.

3. The C-R interaction tends to be stabilized by the addition of detritus relative to the purely community module (no recycling) because the detritus tends to fall out of phase with the resource-nutrient interaction. Out of phase means that when the resource is growing on rising nutrients, the detrital pool is low and recycles nutrients only weakly to the nutrient pool. This reduces resource growth relative to the constant nutrient input assumption of the community module. Similarly, when the resource is declining on declining nutrients, the detrital pool is high and recycles nutrients strongly only to the nutrient pool. This increases resource growth relative to the constant nutrient input assumption of the community module and so prevents it from declining excessively.

4. More theoretical and empirical work is needed in linking food web and ecosystem dynamics.

CHAPTER TWELVE

Food Webs as Complex Adaptive Systems

12.1 SEARCHING FOR EMPIRICAL SIGNATURES

In the preceding chapters, I have attempted to weave together a theory from population level dynamics to whole-ecosystem dynamics. The picture that emerges is one in which an interaction or a subsystem, under certain conditions, acts to mute or excite other interactions or subsystems. After theoretically identifying attributes of food web structures that are stabilizing/ destabilizing, it becomes critical to empirically identify what happens to food web structure when we cross biologically relevant gradients (e.g., ecosystem size). As an example, I argued that spatially expansive ecosystems may be more stable than highly fragmented ecosystems as higher-order mobile consumers may link strongly to many organisms in a spatially constrained habitat. This strong linking in small systems tends to lead to strong top-down pressure and ultimately instability.

If such a fragmented system does inspire tremendous instability, it is likely that the system will not remain in that state for long. Organisms will respond to the changing conditions and so potentially change the underlying food web topology and interaction strength. Further, the instability may drive the loss of species and lower species richness. It is perhaps one of the strongest empirical findings in ecology that species richness scales with area, such that bigger areas hold greater species diversity (Rosenzweig, 1995). Is it possible, at least for islands, lakes, and fragmented habitats, that this reduced diversity is, at a level, a signature of past instability. The increased species extinction rates familiar to MacArthur and Wilson's theory of island biogeography (MacArthur and Wilson, 1967) may be a result of the intensification of interaction strengths in small, isolated ecosystems as well as reduced habitat size.

Thus, the argument is that instability is a transient phenomenon in a complex adaptive food web. Species drop out, others species change whom they consume and how strongly they consume species, and the system eventually attains the ability to persist as an assemblage over ecological time again. This view suggests that only when researchers induce perturbations at the scale of the whole system are they able to actually see the transient instability in action. As a result, our ability to understand instability may require us to look for signatures of past instability.

This notion of looking for signatures of past instability, or surrogates of looming instability, has been discussed recently by researchers investigating the resilience of socioecological systems (Folke et al., 2004; Carpenter et al., 2005). Regime shifts in lakes, for example, have been shown to display evidence of an increase in variance in lake-water phosphorus prior to eutrophication (Carpenter and Brock, 2006). In this chapter, I will follow the lead of this group of researchers to consider some of the potential empirical signatures of instability. If food webs are examples of complex adaptive systems, capable of responding to change, rapid change ought to be potentially driven by rapid adaptive behavioral responses of organisms. Here, I will start by considering the role of adaptive behavior on food web topology and interaction strength. I will then consider the implications of this rapid adaptive behavior for the dynamics and stability of ecosystems. I will end by discussing some of the changing conditions that humans are driving and speculate how these may change ecosystems.

12.2   ADAPTIVE BEHAVIOR, CHANGING FOOD WEB TOPOLOGY, AND ECOSYSTEM SIZE

Much of classic food web ecology has sought to identify similarities in food web structure across all ecosystems [Cohen (1978); Cohen and Newman (1985); Martinez (1991b, 1993); and see Dunne (2006) for an excellent review]. This research attempts to predict the general patterns of topology across all webs. A number of models have been put forth that have been enormously successful at producing patterns akin to those found in our most complete compilation of food webs (Williams, 2000; Stouffer, 2005). Remarkably, the niche model of William and Martinez (2000), a variation on Cohen's earlier cascade model, shows that a single niche dimension produces structural attributes that match food web data. This is an encouraging result for such a complex system as an ecosystem.

While this approach has played a leading role in the development of food web theory, more recently researchers have begun to consider how food webs may be different across ecosystems (Post et al., 2000; Post and Takimoto, 2007; Eveleigh et al., 2007). Although, see Schoener (1989) for an earlier example of this across-system approach for island food webs. It is also valuable to consider how the same web (i.e., basically the same set of players)

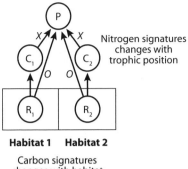

Nitrogen signatures
changes with
trophic position

Habitat 1   Habitat 2

Carbon signatures
changes with habitat

FIGURE 12.1. Depiction of model system. Two habitats form the basis for two pathways that are ultimately coupled by mobile top predators. Note that this mobile top predator is flexible in this system in that it has a choice between habitats (X) and trophic levels (O). Importantly, the coupling aspect of food web adaptability can be empirically ascertained using carbon isotopes when habitats have different carbon signatures, which they often do, while the amount of omnivory can be determined by nitrogen isotopes.

changes across a gradient of factors in space and time (Kondoh, 2003; Krivan and Schmitz, 2004; Beckerman et al., 2006; Petchey et al., 2008). This latter approach of concentrating on differences across ecosystems is closely aligned with another historical food web problem, that of determining what controls food chain length [Pimm (1987); Aunapuu et al. (2008); and see Post (2002) for a good review]. However, the general problem of changing food web topology is not limited to variation in food chain length across ecosystems but can be considered for all food web attributes in space and time.

Let me start by considering the behaviors of those species in food webs that couple multiple habitats in space and the attributes of species that alter food chain length (i.e., omnivory). Adaptive behavioral responses of this sort can be determined empirically and so ought to leave signatures of changing food web topology and strength across a gradient of food webs. A great deal of flexibility in the food web structure will play out in this fashion [the web can expand and contract both horizontally (coupling in space) and vertically (food chain length)].

Tyler Tunney, Brian Shuter, Nigel Lester, and I (Tunney et al., forthcoming), theoretically approached this problem by employing a spatial model similar to that introduced by McCann et al. (2005) and reviewed in chapter 9. Recall that this complete model includes flexibility in both coupling and omnivory (figure 12.1) [see Table 1 in (McCann et al., 2005)]. Luckily, these same architectural attributes (i.e., degree of habitat coupling and omnivory) are relatively easy to get for real food webs using stable isotope approaches. As an example, figure 12.1 highlights that different macrohabitats often have different carbon signatures, allowing researchers to delineate the amount of coupling, while nitrogen isotopes allow us to determine trophic position and so enable an estimate of omnivory, or lower-level feeding. Importantly, the theory and the empirical tools are aligned and so facilitate testing.

The model employed is not exactly an optimal foraging model. Rather, it invokes rapid density-dependent choice by consumers. Thus, consumption qualitatively mimics optimal foraging; however, it is sloppy. There is an advantage to this slightly sloppy foraging approach in that the perfectly optimal is not seen in nature and optimal foraging assumptions have the unfortunate mathematical artifact that the equations are discontinuous and so promote local instability (Fryxell and Lundberg, 1998; Abrams, 2010).

This model allows us to explore how the food web structure responds to changes in ecosystem size and shape. We assumed that a reduced ecosystem reduces the two major habitats similarly (the proportion of each remains the same: just the size is reduced). To keep consistent with the biology of the top predator, lake trout, we assumed that the top predator prefers one habitat (e.g., say the cold-water pelagic habitat). In these model experiments we assume that the scale of foraging of all the predators does not change when we change ecosystem size or shape. The foraging scales of the consumers are assumed to be fixed biological traits. As such, reduced ecosystem size increases the consumer's ability to access prey in both habitats.

To highlight this idea, imagine two conveyor belts that carry different types of prey (one belt might even have preferred prey). If we first consider these conveyor belts in a spatially expansive world, any consumer can forage on only one conveyor belt at a time (there being a fairly long travel time between conveyor belts). As we begin to move these belts together, the consumer is able to move more efficiently back and forth between the two. The consumer therefore begins to increase its overall consumption rates. At the extreme (i.e., a spatially compressed ecosystem), the conveyor belts lie side by side and the consumer, which retains its foraging scale, can ultimately forage on both belts simultaneously. While this is a simple thought experiment, it does highlight that in a spatially restrictive world the consumer may not functionally perceive the spatial structure. This is likely to be especially true when resource heterogeneity operates at a spatial scale smaller than the efficient foraging scale of the consumer.

With this in mind, we modified the spatial scale and examined how the food web topology changed in terms of the habitat coupling strength (X), omnivory (O), food chain length, and the biomass structure of the whole system. This

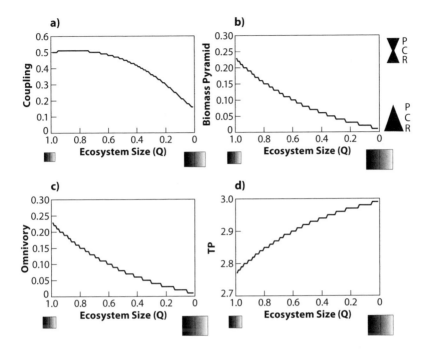

FIGURE 12.2. Prediction of the model system as the ecosystem size is reduced. (a) The amount of prey from the less preferred habitat that the top predator consumes as the system gets smaller. Note an increasing trend. (b) The biomass of consumers relative to resources as the system gets smaller. Note a decreasing trend. (c) The amount of omnivory as the ecosystem size decreases. Omnivory is predicted to increase. (d) The food chain length decreases with decreasing ecosystem size following the omnivory result.

final aspect of this structure determines whether there is lots of biomass high in the trophic structure relative to the lower trophic levels. This is an indicator of biomass flux, as it is well known that increased vertical flux through the web implies stronger top-down control (deBruyn et al., 2007). Further, strong top-down control, or high-energy flux through C-R interactions, tends to correspond to heightened instability [as shown in McCann et al. (2005) and discussed in this book].

As we reduce ecosystem size in this model, it is clear that the food web adapts in a specific way. First, and not surprisingly, the predator increases its coupling into the two distinct habitats (figure 12.2a). This increased coupling ultimately translates into increased trophic flux up to the top predator and so makes the biomass pyramids less Eltonian as space is constrained. In some cases, under extremely high vertical biomass flux up the food web, the biomass pyramid can even take on a wasp-waisted shape (inflated top predator, reduced consumer and released resource biomass) (figure 12.2b). These are our first two predictions of the adaptive driven theory.

This suppression of the intermediate consumer cascades down the trophic structure to free up the resource to flourish. This, therefore, simultaneously makes omnivory tempting for the top consumer (an increased amount of R relative to C). Not surprisingly then, the model system adapts not only by increasing habitat coupling with reduced ecosystem size but also by increasing omnivory (figure 12.2c). Thus, the food chain length is predicted to decrease with decreases in ecosystem size (figure 12.2d). These are our final two predictions. The food web, therefore, widens and truncates its trophic position as ecosystem size decreases and then thins out (reduces spatial coupling) and increases in trophic position with spatially expansive ecosystems. While the theory here has been considered across ecosystem size, such adaptive food web responses may also occur within a given food web as a response to, say, changing productivity in the distinct habitats. Any change in habitat production would be expected to modify the distribution of coupling strengths by the top predator, while a change in the R:C ratio may be expected to alter the amount of omnivory. Again, the food webs should expand and contract in response to the changing abiotic or biotic conditions.

12.3   EMPIRICAL RESULTS FOR CANADIAN SHIELD LAKE ECOSYSTEMS

With these four behaviorally driven predictions for food webs across a range of ecosystem sizes, we then turned to mounting data taken from some empirical investigations of Canadian Shield lake trout food webs. This system is amenable to the above theory because these lakes tend to draw from the same set of species (they draw from relict flora and fauna, although novel species clearly arise) and under a relatively limited range in productivity (i.e., most of the lakes are oligotrophic). As such, they enable a natural experiment across a given set of conditions.

Further, and consistent with the above model, this food web has as a top predator, lake trout, that potentially couple two relatively distinct habitats: (i) the cold-water pelagic zone where they tend to prey on pelagic forage fish like cisco, and (ii) the warm-water littoral zone where they forage on littoral fish and invertebrates (Zanden and Rasmussen, 1999; Vander Zanden and Rasmussen, 2001). The model above, at least as a gross caracature, mimics the food web of these relict lake ecosystems.

In this ecosystem, lake trout are cold-water fish, and the summer water conditions in the southern range of these lake ecosystems has an epilimnion (i.e., the top layer in a stratified lake) that is much warmer than the cold deep-water zone. As a result, the littoral zone becomes a more costly zone thermally for the lake trout, and so the predatory lake trout tend to prefer the cold pelagic zone at this time of year. Nonetheless, researchers have found that they make 10 to 15-minute forays into the near-shore zone (Morbey et al., 2006). The size of this near-shore zone, therefore, can play a huge role in how this food web is constructed. For example, a very large littoral near-shore zone may mean that near-shore prey fish can move away from the cold-water zone occupied by the lake trout. The lake trout simply have little chance of finding the prey as they make their limited ventures into the warm-water zone. In big enough littoral zones, then, the outer portions of the lake may be refugia from these top predators.

Other species are similarly afflicted by the thermal boundaries. Cisco, or lake herring, are cold-water zooplanktivorous fish that lake trout feed upon in the deep cold-water regions of a lake. Like the trout, the cisco are somewhat thermally constrained to the cold-water zone in the summer. Other prey fish for the trout, like perch and minnow, tend to live in the near-shore habitats, feeding on zooplankton and benthic invertebrates. Thus, the food web is relatively spatially distinct, with a near-shore warm-water web and a deep cold-water pelagic web.

Given all this, it is natural then to consider how the food web for this Canadian Shield lake ecosystem changes across ecosystem size. Does this web expand horizontally (i.e., couple into the littoral) and contract vertically (i.e., increase the amount of omnivory and decrease the food chain length) as lake size decreases? We will also consider the related case of what happens to the food web when the lake's shape (i.e., the relative amount of littoral habitat to pelagic habitat) is altered.

A number of researchers have been gathering information on this classic Canadian lake ecosystem (Cabana and Rasmussen, 1994; Zanden and Rasmussen, 1999; Gunn et al., eds, 2004; Dolson et al., 2009), and through the use of stable isotopes and stomach contents, we are now in a position to consider the predictions from the above theory. As predicted above, these shield lakes show increased coupling (figure 12.3a), increased top predator (lake trout) density (figure 12.3b), increased omnivory (figure 12.3c), and decreased trophic position (figure 12.3d) with decreased ecosystem size. However, the reasons for increased omnivory are not clear (Post et al., 2000; Post, 2002; McCann et al., 2005), as we do not have data on the community abundances across this same gradient. The lake trout density increases

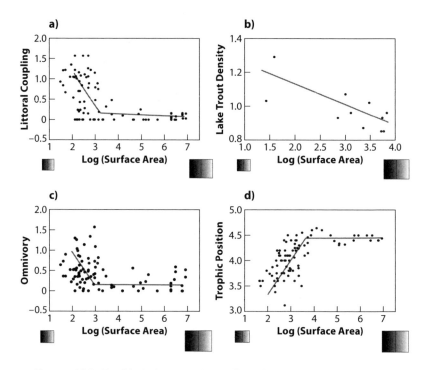

FIGURE 12.3. Empirical data on (a) coupling, (b) top predator density, (c) omnivory, and (d) food chain length for lake trout food webs in the Canadian Shield.

as predicted, but it is unclear whether prey fish are suppressed as predicted. Other data gathered from different lake ecosystems suggest that this pattern may also be correct. For example, a survey of numerous lakes by Helene Cyr and Rob Peters found that indeed the biomass resource spectrum (which presumably correlates with the trophic position) increases in slope as the lake size decreases (Cyr and Peters, 1996). Such a change ought to promote increased omnivory as discussed above.

Recognizing this interesting pattern, Rebbecca Dolson, Neil Rooney, Mark Ridgway, and I decided to further examine this lake trout food web response (Dolson et al., 2009). But this time we attempted to minimize the range of the lake's size and instead concentrated on altering the relative amount of littoral habitat. In the lakes we studied in Algonquin Park, several hours north of Toronto, this relative change in lake habitat types tended to correlate reasonably well with lake shape. Circular bowl-shaped lakes tended to have small relative littoral zone sizes, while reticulate complex lakes had relatively more shallow littoral habitat.

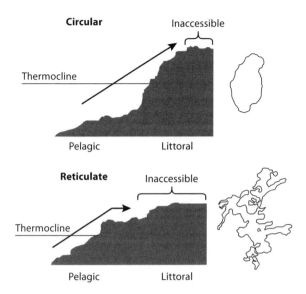

FIGURE 12.4. Schematic of how lake trout foraging in warm-water zones may be influenced by the size and shape of the lake. Bowl-shaped lakes, in this case associated with small littoral zones, allow easy access to littoral habitats, while lakes with extensive littoral habitats are not as easily accessed because of the thermal constraints acting on lake trout. In this study, circular lakes tended to be associated with bowl-shaped lakes and small littoral zones, however, this pattern does not necessarily hold. The main point is that lakes in which all littoral habitats are "everywhere near," the cold-water habitat makes for an easily accessible littoral habitat for the trout. From Dolson et al. (2009). © 2005. John Wiley & Sons.

Given lakes with a smaller amount of littoral zone relative to the cold deep-water zone, all littoral prey should be relatively close to the cold-water lake trout habitat. That is, a lake trout foraging for 15 minutes potentially had access to all the habitat during any near-shore foraging event. On the other hand, a lake trout foraging in a spatially expansive near-shore zone would have many areas it could never reach in the summer. This empirical survey of different lake shapes thus presented us with another case to test the behaviorally driven predictions discussed above. Again, one might expect that circular lakes (where the littoral zone is everywhere near a lake trout residing at the edge of the deep cold-water zone) to show heightened coupling (figure 12.4). Further, this heightened access to littoral prey, at reduced cost, ought to enhance the flux of energy up the food web, and so top-down control should readily materialize with increased littoral access. In a sense, I am suggesting that the

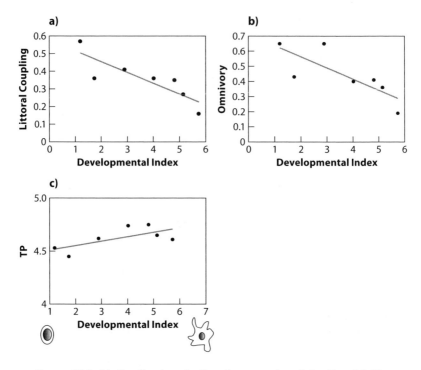

FIGURE 12.5. (a) Coupling into the littoral zone as the relative littoral habitat changes (here, the lake developmental index is correlated with the relative littoral habitat (Dolson et al., 2009)). (b) Changes in omnivory as the relative littoral habitat changes. (c) Changes in food chain length as the relative littoral habitat changes. Patterns again match model predictions.

reduced relative littoral zone size should behave similarly to the reduced lake size, which also naturally reduces the littoral zone size.

Although we were able to census only seven lakes, the pattern that emerges is consistent and significant. Again, the habitat coupling by lake trout increased with increased access to the nonpreferred littoral habitat (figure 12.5), and the amount of omnivory in both channels increased with a reduced relative littoral zone (figure 12.5). Thus, we found a reduced food chain length with reduced littoral zones (figure 12.5e). More data on this pattern will be important to verify the regularity of this result, whereas the lake size effect has been thoroughly examined and repeated.

In summary, the food web changes consistently across both lake size and shape, as depicted in figure 12.6. The food web expands horizontally by coupling more into littoral zone prey when lakes are small (or littorally small) and simultaneously decreases in food chain length. In a sense, the individual

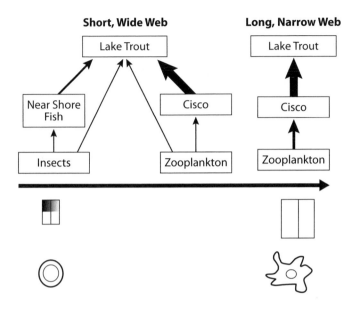

FIGURE 12.6. Schematic of how lake trout food webs change with the size and shape of the lake. The web expands and contracts like an accordion. Lakes truncate and widen when the warm littoral zone is everywhere near The cold pelagic zone and elongate and narrow for lakes with extensive warm-water littoral areas.

behavior causes the food web to adapt by expanding and contracting in terms of trophic position. Although beyond the scope of this book, it appears that the life history attributes of the lake trout change along this gradient consistently as well. Tyler Tunney, Brian Shuter, Nigel Lester, and I have found in both cases so far that lake trout tend to be much larger and mature later in big lakes (this also appears relatively true in big littoral lakes) than in small lakes, where lake trout are quite small and mature earlier. Some of this change appears also due in part to changes in the composition of the food web. As an example, lake trout without cisco tend to be much smaller on average (Martin, 1966; Rasmussen et al., 1990).

I have argued with the theory that topology, interaction strength, and community structure (i.e., biomass pyramids) likely change across fundamental gradients (e.g., ecosystem size and shape). However, I have ignored the fact that along this same gradient the system may also lose species because of realized "instabilities" of the strong kind, local extinction. The theory suggests that this is a real possibility, as the heightened coupling into the two channels increases production up to the predator and in turn drives increased suppression of intermediate consumers. This does two things: It drives potentially

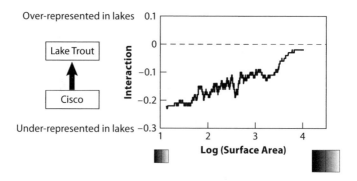

FIGURE 12.7. Expected versus actual frequency of lakes with both cisco and lake trout. These results suggest that among small lakes the number of lakes with both cisco and lake trout are underrepresented, while among larger lakes the actual number of lakes with both cisco and lake trout approach the number of lakes expected. Expected and realized representations were taken as a running average over a window size of 20 lakes.

fatal suppression of both the littoral (e.g., littoral minnow) and pelagic prey (e.g., cisco) of lake trout and possible periods of boom-and-bust dynamics [see McCann et al. (2005) for a full explanation of this].

Thus, the system may also change by losing species. Since cisco appears to be the lake trout's preferred prey, it is likely that this interaction can be especially intense. Thus, it is possible that these two species cannot persist in small lakes if the interaction gets too strong. This is tricky to approach empirically, but Nigel Lester of the Ontario Ministry of Natural Resources gave me access to an enormous database on fish assemblage diversity in hundreds of Canadian Shield lakes. From this database we can examine the expected and realized frequencies of cisco–lake trout pairings in these food webs. Let me be more clear about this. I took a lake size range (e.g., 100–200 ha) and calculated the percentage of lakes that had cisco, p_C, the percentage of lakes that had lake trout, p_{LT}, and the percentage of lakes that had both together, $p_C - p_{LT}$. I then calculated the expected number of lakes—if the world operates by chance alone—that should have both lake trout and cisco together, Expected$(p_C - p_{LT}) = p_C * p_{LT}$. With all this information in tow, I then plotted the expected and actual proportions of lakes with both lake trout and cisco as a function of lake size (figure 12.7).

If lake trout are able to exert strong top-down suppression on cisco as lake size decreases, one would expect to find that as lake size decreases, the number of lakes that have both species together should be less than expected by chance alone. This is indeed true: In large lakes both are present as expected, while in small lakes we find that lakes with both lake trout and cisco are underrepresented. This is not proof that small lakes excacerbate instability, but it is consistent—figure 12.7 may therefore represent a signature of past instabilities. Small lakes also have a lower diversity of littoral prey as well; however, again this may be because of a multitude of other factors (less niche space, etc.). Nonetheless, some aspect of this loss of littoral prey diversity may also be a signature of past food web instabilities.

In this chapter, we have begun to see that some of the patterns in diversity, topology, and interaction strength may be due to the changing nature of real-world food webs. That is, organisms adapt to the conditions of their environment behaviorally, ultimately in an evolutionarily sense, and in doing so change the topology and interaction strength of food webs. Further, some of the changes in food webs of a given type, say boreal lake ecosystems, may also be due to the idea that food web instabilities potentially create historical extinctions. Small ecosystems have the potential for increased extinction rates because of stronger interactions, and this increased vulnerability to extinction is consistent with the assumptions of MacArthur and Wilson's theory of island biogeography. Of course, many factors may contribute to this increased vulnerability with reduced ecosystem size.

While I have focused on lake ecosystems, it is worth pointing out that this same set of arguments can be used for island ecosystems. Further, the results here suggest that fragmented ecosystems, which reduce ecosystem size, can be expected to also undergo similar changes with accompanying losses in diversity (Terborgh et al., 2001). The above results suggest that a metric of top-down control (e.g., a biomass pyramid) may also be a good signature of the potential for instability. An inverted pyramid suggests runaway growth in the upper trophic levels. We now consider how human influences and ecosystems coupled in space may drive such biomass pyramids, potentially leading to species loss.

12.4 SUBSIDIES, OPPORTUNISTS, AND HOMOGENIZATION

Gary Polis championed the view that most systems rely, in some way or another, on a subsidy from another system (Polis and Strong, 1996; Polis and Hurd, 1996; Polis et al., 1997). Polis and Strong's (1996) seminal paper used empirically motivated arguments to point out that focal food chains often are structured strongly by an "external" contribution. This viewpoint of subsidies playing a major and underappreciated role likely arose naturally for Gary Polis while he studied desert islands surrounded by the enormously productive Sea of Cortez, an area known for its strong upwelling. These islands are swamped regularly with life from the productive nearby sea. Additionally, marine bird populations that nest on these islands forage from the sea only to return to the desert islands and deposit ocean-rich nutrient-filled guana. The community structure and production of this island therefore depend dramatically on ocean subsidies (Stapp and Polis, 2003). Organisms that can take advantage of the subsidy stand to outcompete organisms deriving energy solely from the insular part of the island ecosystem. In a synthetic contribution, Leroux and Loreau (2008) extended these ideas in a simple, elegant paper to suggest that the influence subsidies have on specific ecosystems can be understood from a landscape perspective. While subsidies exist everywhere, their influence likely spans a range of strength.

The prevalence of subsidies may be taking on an even larger role with the increasing influence of humans. Numerous examples exist where human influences act as a subsidy that drives enormous changes in ecological communities. Sometimes these subsidies cause whole changes in the production of a major ecosystem (Dodds, 2006). In many cases, an organism capable of harnessing human-driven subsidies flourishes and suddenly dominates a whole ecosystem (Jefferies et al., 2006). It may be that extremely generalized organisms, for example, are the most likely to captialize on these unnatural subsidies and rise to high densities to wreak havoc on whole ecosystems.

The sea urchin example in chapter 1 is an example of this type of phenomenon where excessive nutrient runoff near coastal regions elevates productivity that the generalized urchin harnesses (Tewfik et al., 2007b). In turn, the urchin mows down edible seagrasses and consumes all the detritus. The benthic component of the system is suppressed, and the more natural multichannel functioning appears to become largely propelled by pelagic production.

Some of the above outcomes of human change result in a functional reduction in habitats (e.g., urchins mow down grasses and detritus). It seems that the influence of humans has some commonalities. One of these is that such a massive destructive force tends to homogenize natural ecosytems. Where there once were fields and forests now there are cities and agriculture, and so the diversity of habitats is reduced considerably. These cities and agricultural developments produce lopsided amounts of nutrients that run off into adjacent ecosystems, potentially overwhelming existing pathways of energy and turning the systems into one dominated by a single human-driven subsidy.

The complexity of energy pathways forming food webs and whole ecosystems is thus greatly diminished. The domination of a single pathway in a once-reticulate web ought to increase the mean interaction strength driving strong, potentially damaging, top-down cascading influences on whole communities. Individuals behave, though, and through these means systems change as discussed in this chapter. It remains to be seen how much this homogenization means to ecosystem stability and function. One would believe we have already heavily eroded critical structure. Recognize here that strong and consistent subsidies can drive a dynamically stable consumer. However, this same constant flux potentially allows this dynamically stable consumer (low variability) to suppress prey to local extinction (the ultimate instability). As discussed throughout, this dynamical stability makes sense because the consumer is decoupled from its energy source. The local extinction of prey also is inevitable given such artifically inflated inverted biomass pyramids.

I have looked at food webs in space via multiscale foraging, but I have not really spoken to how dispersal influences food web dynamics. There is no doubt it likely plays a major role in dynamics. Gouhier et al. (2010) have shown recently that the metacommunity concept is likely to influence stability because dispersal, environmental variability, and compensatory dynamics clearly interact. It remains of fundamental importance to further theoretically and empirically decipher how both dispersal and physical coupling at all trophic levels influence interspecific asynchrony and ultimately food web dynamics.

12.5   HUMANS IN THE FOOD WEB

Ecosystems are currently losing species at unprecedented rates. Recognizing this, Ives and Cardinale (2004) considered how the order of species extinction might influence the stability of food webs. Because compensatory food web responses enable a system to manage its response to environmental changes, Ives and Cardinale (2004) pointed out with simulations that the loss of certain species can make a community less able to withstand further environmental degradation. They used their results to point out that the order of extinction may also lead to a once-minor species becoming a major food web player in the future. In the face of relatively unpredictable future perturbations, it is arguable that all species, even minor players, may end up being of major importance. Further, I have argued here that food webs ignore ecosystems and habitat divisions, as couplers move between these human-defined boundaries without concern. This movement may be critical to the sustainability of these systems, and so it may be wise to consider landscape-scale management practices instead of focusing on charismatic single species.

Here, and throughout the book, I have argued that behavior plays a important role in the dynamical response of systems to variation. Human behavior may also play a role in the dynamics of exploited resources. Oddly, much resource management theory has assumed constant quotas or harvests and ignored the behavioral dynamics of harvesters/foragers. Humans have often been considered decoupled from the dynamics, perhaps because we are such generalists and our population dynamics no longer wax and wane in response to resource fluctuations or declines. However, human effort dynamics may follow population fluctuations of resources, and so humans may still be the dynamical drivers of resources. Recently, for example, Fryxell et al. (2010) re-examined data from a number of harvested populations and looked for dynamical responses of harvesters to changing population densities of the resource.

In all cases, data for the change in effort increased positively with increases in resource density, while the change in effort responded negatively to increases in effort. Fryxell et al. (2010) derived a simple model that incorporated this density-dependent effort response and found that this behavior drove classic consumer-resource quasicycles or cycles. Further, they found empirical evidence that this weak compensatory response by humans to changing levels of resource abundance actually appears to induce harvest cycles (figure 12.8).

The effort dynamics act much like traditional consumer-resource models, with consumer response (effort and quotas) lagging behind resource dynamics. Dynamical system models incorporating this mix of feedback predict that cycles or quasicycles with decadal periodicity should commonly occur in harvested wildlife populations, with effort and quotas lagging behind resources, whereas harvests should exhibit less lagged responses. Empirical data gathered from three hunted populations of white-tailed deer and moose were consistent with these predictions of both underlying behavioral causes and dynamical consequences (figure 12.8). Figure 12.8a shows the predictions of the model to stochastic perturbation, while the three natural populations all follow the lagged predictions discussed above.

Humans operate at the whole-landscape scale. While a huge spatial scale, it may be that all ecosystems are spatially constrained for humans. Human innovation is a powerful force, and technology has allowed them to extract resources from the environment with unbelievable efficiency across space and time. This efficiency, spread across ecosystems, may mean that human impacts act to synchronize the dynamics of resources on the landscape. The world has become small for humans, and so switching responses do not play out when they are the top predator. At a level, human activity is attacking ecosystems from the top and bottom simultaneously. The culling of major resources removes or depletes species from the top of the food chain to the point of reducing empirical estimates of food chain lengths in many ecosystems

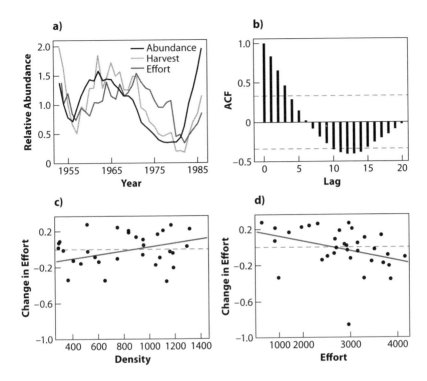

FIGURE 12.8. Harvest data are illustrated for white-tailed deer from the Canonto District of Ontario, Canada. Panel A shows annual estimates of resource abundance (filled circles), harvest (open circles), and effort (filled triangles). To simplify plotting on a single set of axes, variables were normalized by dividing yearly values by the mean for the entire time series. Autocorrelation functions for resource abundance are shown in panel B. All correlation coefficients that exceed the horizontal bars are statistically significant (Bartlett's test, $p < 0.05$). The bottom two plots show rates of change in hunting effort (z) in relation to resource density (x, panel C) and current effort (y, panel D) ($z = 0.0706 + 0.000305x - 0.000104y, F_{2,30} = 5.51, p = 0.009, R^2 = 0.271$). © 2010. The American Association for the Advancement of Science.

(Pauly et al., 1998; Essington et al., 2006). At the same time, nutrient loading and the alteration of major nutrient cycles on the landscape appear to often have the consequence of empowering a single energy pathway in a food web and limiting the production of other important pathways. Further, habitat fragmentation is steadily reducing the average size of ecosystems and increasing the distance between ecosystems, while habitat modification homogenizes once structurally replete ecosystems. If the structure of natural ecosystems is important to their stability and sustainability, the ecological collapses discussed in chapter 1 may be only just beginning.

Bibliography

Abraham, R., and C. D. Shaw. 1982. *Dynamics: the geometry of behavior.* Aerial Press, Santa Cruz, CA.

Abrams, P. A. 1999. "The adaptive dynamics of consumer choice." *American Naturalist* 153(1):83–97.

Abrams, P. A. 2007. "Habitat choice in predator-prey systems; spatial instability due to interacting adaptive movements." *American Naturalist* 169:581–594.

Abrams, P. A. 2010. "Quantitative descriptions of resource choice in ecological models." *Population Ecology* 52:47–58.

Allesina, S., and M. Pascual. 2008. "Network structure, predator–prey modules, and stability in large food webs." *Theoretical Ecology* 1(1):55–64.

Anderson, C.N.K., C. Hsieh, S. A. Sandin, R. Hewitt, A. Hollowed, J. R. Beddington, R. M. May, and G. Sugihara. 2008. "Why fishing magnifies fluctuations in fish abundance." *Nature* 452:835–839.

Andersson, J., P. Bystrom, D. Claessen, L. Persson, and A. M. De Roos. 2007. "Stabilization of population fluctuations due to cannibalism promotes resource polymorphism in fish." *American Naturalist* 169(6):820–829.

Arim, M., S. R. Abades, G. Laufer, M. Loureiro, and P. A. Marquet. 2010. "Food web structure and body size: trophic position and resource acquisition." *Oikos* 119:147–153.

Artzy-Randrup, Y., and L. Stone. 2004. "Comment on 'Network motifs: simple building blocks of complex networks' and 'Superfamilies of evovled and designed networks'." *Science* 305:1107–1108.

Assessment, M. E. 2005. *Ecosystems and human well-being: wetlands and water.* World Resources Institute, Washington, DC.

Aunapuu, M., J. Dahlgren, T. Oksanen, D. Grellmann, L. Oksanen, J. Olofsson, U. Rammul, M. Schneider, B. Johansen, and H. O. Hygen. 2008. "Spatial patterns and dynamic responses of arctic food webs corroborate the Exploitation Ecosystems Hypothesis (EEH)." *American Naturalist* 171:249–262.

Barrow-Green, J. 1997. *Poincaré and the three body problem. History of Mathematics, vol. II.* American Mathematical Society, Providence, RI.

Bascompte, J., and C. J. Melian. 2005. "Trophic modules for complex food webs." *Ecology* 86:2868–2873.

Bascompte, J., C. J. Melian, and E. Sala. 2005. "Interaction strength combinations and the overfishing of a marine food web." *Proceedings of the National Academy of Sciences of the United States of America* 102(15):5443–5447.

Beckerman, A. P., O. L. Petchey, and P. H. Warren. 2006. "Foraging biology predicts food web complexity." *Proceedings of the National Academy of Sciences of the United States of America* 103(37):13745–13749.

Bellman, R., and K. L. Cooke. 1963. *Differential-difference equations.* Academic Press, New York.

Berberian, S. K. 1992. *Linear algebra.* Oxford Science Publications, Oxford.

Berlow, E. L., A. M. Neutel, J. E. Cohen, P. C. de Ruiter, B. Ebenman, M. Emmerson, J. W. Fox, V.A.A. Jansen, J. I. Jones, G. D. Kokkoris, D. O. Logofet, A. J. McKane, J. M. Montoya, and O. Petchey. 2004. "Interaction strengths in food webs: issues and opportunities." *Journal of Animal Ecology* 73(3):585–598.

Brose, U., T. Jonsson, E. L. Berlow, P. Warren, C. Banasek-Richter, L. F. Bersier, J. L. Blanchard, T. Brey, S. R. Carpenter, M.F.C. Blandenier, L. Cushing, H. A. Dawah, T. Dell, F. Edwards, S. Harper-Smith, U. Jacob, M. E. Ledger, N. D. Martinez, J. Memmott, K. Mintenbeck, J. K. Pinnegar, B. C. Rall, T. S. Rayner, D. C. Reuman, L. Ruess, W. Ulrich, R. J. Williams, G. Woodward, and J. E. Cohen. 2006a. "Consumer-resource body-size relationships in natural food webs." *Ecology* 87(10): 2411–2417.

Brose, U., R. J. Williams, and N. D. Martinez. 2006b. "Allometric scaling enhances stability in complex food webs." *Ecology Letters* 9(11):1228–1236.

Brown, J. H., J. F. Gillooly, A. P. Allen, V. M. Savage, and G. B. West. 2004. "Toward a metabolic theory of ecology." *Ecology* 85(7):1771–1789.

Cabana, G., and J. B. Rasmussen. 1994. "Modelling food-chain structure and contaminant bioaccumulation using stable nitrogen isotopes." *Nature* 372:17.

Camacho, J., D. B. Stouffer, and L. A. N. Amaral. 2007. "Quantitative analysis of the local structure of food webs." *Journal of Theoretical Biology* 246:260–268.

Cantrell, R. S., and C. Cosner. 2001. "On the dynamics of predator-prey models with the Beddington-DeAngelis functional response." *Journal of Mathematical Analysis and Applications* 257:206–222.

Cappuccino, N., D. Lavertu, Y. Bergeron, and J. Regniere. 1998. "Spruce budworm impact, abundance and parasitism rate in a patchy landscape." *Oecologia* 114(2):236–242.

Carnicer, J., P. A. Abrams, and P. Jordano. 2008. "Switching behavior, coexistence and diversification: comparing empirical community-wide evidence with theoretical predictions." *Ecology Letters* 11(8):802–808.

Carpenter, S. R., and W. A. Brock. 2006. "Variance: a leading indicator of ecological transition." *Ecology Letters* 9:311–318.

Carpenter, S. R., and J. F. Kitchell. 1996. *The trophic cascade in lakes.* Cambridge University Press, Cambridge.

Carpenter, S. R., F. Westley, and M. G. Turner. 2005. "Surrogates for resilience of social-ecological systems." *Ecosystems* 8:941–944.

Carson, R. 1962. *Silent spring.* Houghton Mifflin. Boston.

Cattin, M. F. 2004. "Phylogenetic constraints and adaptation explain food web structure." *Nature* 427:835–839.

Cebrian, J., J. B. Shurin, E. T. Borer, B. J. J. Cardinale, T. Ngai, M. D. Smith, and W. F. Fagan. 2009. "Producer nutritional quality controls ecosystem trophic structure." *PloS One* 4:e4929.

Charnov, E. L. 1976. "Optimal foraging: marginal value theorem." *Theoretical Population Biology* 9(2):129–136.

Chesson, J. 1983. "The estimation and analysis of preference and its relationship to foraging models." *Ecology* 64(5):1297–1304.

Claessen D., A. M. De Roos, and L. Persson. 2000. "Dwarfs and giants: cannibalism and competition in size-structured populations." *American Naturalist* 155(2):219–237.

———. 2004. "Population dynamic theory of size-dependent cannibalism." *Proceedings of the Royal Society of London Series B: Biological Sciences* 271(1537):333.

Clauset, A., C. Moore, and M.E.J. Newman. 2008. "Hierarchical structure and the prediction of missing links in networks." *Nature* 453(7191):98–101.

Cohen, J. E. 1978. *Food webs and niche space*. Princeton University Press, Princeton, NJ.

Cohen, J., F. Briand, and C. M. Newman. 1990. *Food webs and niche space. Monographs in Population Biology 11*. Springer, New York.

Cohen, J. E., and C. M. Newman. 1985. "A stochastic-theory of community food webs. 1. Models and aggregated data." *Proceedings of the Royal Society of London Series B: Biological Sciences*, 224(1237):421–448.

Cohen, J. E., C. M. Newman, and F. Briand. 1985. "A stochastic theory of community food webs. 2. Individual Webs," *Proceedings of the Royal Society of London Series B: Biological Sciences* 224(1237):449–461.

Comins, H. N., M. P. Hassell, and R. M. May. 1992. "The spatial dynamics of host parasitoid systems." *Journal of Animal Ecology* 61(3):735–748.

Costantino, R. F., R. A. Desharnais, J. M. Cushing, and B. Dennis. 1997. "Chaotic dynamics in an insect population." *Science* 275(5298):389.

Costanza, R., R. d'Arge, R. deGroot, S. Farber, M. Grasso, B. Hannon, K. Limburg, S. Naeem, R. V. O'Neill, J. Paruelo, R. G. Raskin, P. Sutton, and M. van den Belt. 1997. "The value of the world's ecosystem services and natural capital." *Nature* 387(6630):253–260.

Crowley, P. H. 1981. "Dispersal and the stability of predator-prey interactions." *American Naturalist*. 118:673–701.

Cyr, H., and M. L. Pace. 1993. "Magnitude and patterns of herbivory in aquatic and terrestrial ecosystems." *Nature* 361:148–150.

Cyr, H., and R. H. Peters. 1996. "Biomass-size spectra and the prediction of fish biomass in lakes." *Canadian Journal of Fisheries and Aquatic Sciences* 53(5):994–1006.

Dambacher, J. M., H. K. Luh, H. W. Li, and P. A. Rossignol. 2003. "Qualitative stability and ambiguity in model ecosystems." *American Naturalist* 161:876–888.

DeAngelis, D. L. 1975. "Stability and connectance in a food web model." *Ecology* 56:238–243.

———. 1980. "Energy-flow, nutrient cycling, and ecosystem resilience." *Ecology* 61(4):764–771.

———. 1992. "Dynamics of nutrient cycling and food webs." Chapman and Hall, New York.

DeAngelis, D. L., and R. A. Goldstein. 1978. "Criteria that forbid a large, non-linear food-web model from having more than one equilibrium point." *Mathematical Biosciences* 41(1–2):81–90.

deBruyn, A. M., K. S. McCann, J. C. Moore, and D. R. Strong. 2007. "An energetic framework for trophic control." In: *From energetics to ecosystems*, edited by N. Rooney, K. S. McCann, and D. Noakes, 65–85. Springer, Dordrecht, Netherlands.

deBruyn, A. M., H. M. Trudel, N. Eyding, J. Harding, H. McNally, R. Mountain, C. Orr, D. Urban, S. Verenitch, and A. Mazumder. 2006. "Ecosystemic effects of salmon farming increase mercury contamination in wild fish." *Environmental Science & Technology* 40(11):3489–3493.

de Koppel, J. V., R. D. Bardget, J. Bengstton, C. Rodriguez-Barrueco, M. Rietkirk, M. Wassen, and V. Wolters. 2005. "The effects of spatial scale on trophic interactions." *Ecosystems* 8:801–807.

De Roos, A. M., E. McCauley, and W. G. Wilson. 1998. "Pattern formation and the spatial scale of interaction between predators and their prey." *Theoretical Population Biology* 53(2):108–130.

De Roos, A. M., and L. Persson. 2002. "Size-dependent life-history traits promote catastrophic collapses of top predators." *Proceedings of the National Academy of Sciences of the United States of America* 99(20):12907–12912.

De Roos, A. M., L. Persson, and E. McCauley. 2003. "The influence of size-dependent life-history traits on the structure and dynamics of populations and communities." *Ecology Letters* 6(5):473–487.

de Ruiter, P. C., A. M. Neutel, and J. C. Moore. 1995. "Energetics, patterns of interaction strengths, and stability in real ecosystems." *Science* 269(5228):1257–1260.

Doak, D. F., D. Bigger, E. K. Harding, M. A. Marvier, R. E. O'Malley, and D. Thomson. 1998. "The statistical inevitability of stability-diversity relationships in community ecology." *American Naturalist* 151(3):264–276.

Dobson, A., S. Allesina, K. Lafferty, and M. Pascual. 2009. "The assembly, collapse and restoration of food webs." *Philosophical Transactions of the Royal Society of London Series B: Biological Sciences* 364(1524):1803–1806.

Dodds, W. K. 2006. "Nutrients and the 'dead zone': the link between nutrient ratios and dissolved oxygen, in the northern Gulf of Mexico." *Frontiers in Ecology and the Environment* 4(4):211–217.

Dolson, R., K. S. McCann, N. Rooney, and M. Ridgway. 2009. "Lake shape predicts the degree of habitat coupling by a mobile predator." *Oikos* 118:1230–1238.

Drummund, H. 1983. "Aquatic foraging in garter snakes: a comparison of specialists and generalists." *Behaviour* 86:1–30.

Dunne, J. A. 2006. "The network structure of food webs." In *Ecological networks: linking structure to dynamics in food webs*, edited by M. Pascual, and J. A. Dunne, 27–86. Oxford University Press, Oxford.

Dunne, J. A., R. J. Williams, and N. D. Martinez. 2002. "Food-web structure and network theory: the role of connectance and size." *Proceedings of the National Academy of Sciences of the United States of America* 99(20):12917–12922.

———. 2004. "Network structure and robustness of marine food webs." *Marine Ecology-Progress Series* 273:291–302.

Durrett, R., and S. Levin. 1994. "The importance of being discrete (and spatial)." *Theoretical Population Biology* 46(3):363–394. http://dx.doi.org/10.1006/tpbi.1994.1032.

Ebenhard, T. 1988. "Introduced birds and mammals and their ecological effects." *Swedish Wildlife Research* 13:1–107.

Edwards, A. M. 2001. "Adding detritus to a nutrient-phytoplankton-zooplankton model: a Dynamical-Systems Approach." *Journal of Plankton Research* 23: 389–413.

Elton, C. 1958. *The invasion of animal and plant communities*. Methuen, London.

Emmerson, M., T. M. Bezemer, M. D. Hunter, T. H. Jones, G. J. Masters, and N. M. Van Dam. 2004. "How does global change affect the strength of trophic interactions?" *Basic and Applied Ecology* 5(6):505–514.

Essington, T. E., A. H. Beaudreau, and J. Wiedenmann. 2006. "Fishing through marine food webs." *Proceedings of the National Academy of Sciences of the United States of America* 103(9):3171–3175.

Essington, T. E., and S. Carpenter. 2000. "Mini-review: nutrient cycling in lakes and streams—insights from a comparative analysis." *Ecosystems* 3:131–143.

Eveleigh, E. S., K. S. McCann, P. C. McCarthy, S. J. Pollock, C. J. Lucarotti, B. Morin, G. A. McDougall, D. B. Strongman, J. T. Huber, J. Umbanhowar, and L.D.B. Faria. 2007. "Fluctuations in density of an outbreak species drive diversity cascades in food webs." *Proceedings of the National Academy of Sciences of the United States of America* 104:16976–16981.

Fagan, W. F. 1997. "Omnivory as a stabilizing feature of natural communities." *American Naturalist* 150(5):554–567. http://www.journals.uchicago.edu/doi/abs/10.1086/286081.

Folke, C., S. Carpenter, B. Walker, M. Scheffer, T. Elmqvist, L. Gunderson, and C. S. Holling. 2004. "Regime shifts, resilience, and biodiversity in ecosystem management." *Annual Review of Ecology Evolution and Systematics* 35:557–581.

Forsyth, D. M., and P. Caley. 2006. "Testing the irruptive paradigm of large-herbivore dynamics." *Ecology* 87:297–303.

Fritts, T. H., and G. H. Rodda. 1998. "The role of introduced species in the degradation of island ecosystems: a case history of Guam." *Annual Review of Ecology and Systematics* 29:113–140. http://www.jstor.org/stable/221704.

Fry, B., and E. B. Sherr. 1984. "Delta-C-13 measurements as indicators of carbon flow in marine and fresh-water ecosystems." *Contributions in Marine Science* 27, SEP:13–47.

Fryxell, J. M., and P. Lundberg. 1998. *Individual behavior and community dynamics*. Chapman & Hall, New York.

Fryxell, J. M., A. Mosser, A.R.E. Sinclair, and C. Packer. 2007. "Group formation stabilizes predator-prey dynamics." *Nature* 449:1041–U4.

Fryxell, J. M., C. Pacard, K. S. McCann, E. J. Solberg, and B. E. Saether. 2010. "Resource management cycles and the sustainability of harvested wildlife populations." *Science* 328:903–906.

Fussmann, G. F., and B. Blasius. 2005. "Community response to enrichment is highly sensitive to model structure." *Biology Letters* 1:912.

Fussmann, G. F., S. P. Ellner, K. W. Shertzer, and N. G. Hairston. 2000. "Crossing the Hopf bifurcation in a live predator-prey system." *Science* 290(5495):1358–1360.

Gardner, M. R., and W. R. Ashby. 1970. "Connectance of large dynamic (cybernetic) systems: critical values for stability." *Nature*: 228(5273):784. http://dx.doi.org/10.1038/228784a0.

Gause, G. F. 1932. "Experimental studies on the struggle for existence. I. Mixed populations of two species of yeast." *Journal of Experimental Biology and Fertility of Soils* 9:389.

———. 1934. *Struggle for existence*. Hafner, New York.

Gellner, G., and K. S. McCann. Forthcoming. "Reconciling the omnivory-stability debate." *American Naturalist*.

Gouhier, T. C., F. Guichard, and A. Gonzalez. 2010. "Synchrony and stability of food webs in metacommunities." *American Naturalist* 175:E16–34.

Gross, T., L. Rudolf, S. A. Levin, and U. Dieckmann. 2009. "Generalized models reveal stabilizing factors in food webs." *Science* 325(5941):747.

Guckenheimer, J., and P. Holmes. 1983. *Nonlinear oscillations, dynamical systems, and bifurcations of vector fields.* Springer, New York.

Gunn, J., R. J. Steedman, and R. Ryder, eds. 2004. *Boreal Shield watersheds' lake trout ecosystems in a changing environment.* CRC Press, Lewis Publishers, Boca Raton, FL.

Hairston, N. G. Jr., and N. G. Hairston Sr. 1993. "Cause-effect relationships in energy-flow, trophic structure, and interspecific interactions." *American Naturalist* 142(3):379–411.

Hanski, I., L. Hansson, and H. Henttonen. 1991. "Specialist predators, generalist predators, and the microtine rodent cycle." *Journal of Animal Ecology* 60:353–367.

Hassell, M. P. 1978. *The dynamics of arthropod predator-prey systems.* Princeton University Press, Princeton, NJ.

Hastings, A. 1983. "Age-dependent predation is not a simple process. 1. Continuous-time models." *Theoretical Population Biology*, 23(3): 347–362.

Hastings, A., and R. F. Costantino. 1987. "Cannibalistic egg-larva interactions in *Tribolium*: an explanation for the oscillations in population numbers." *American Naturalist*, 130(1):36–52.

Hastings, A., S. Harrison, and K. McCann. 1997. "Unexpected spatial patterns in an insect outbreak match a predator diffusion model." *Proceedings of the Royal Society of London Series B: Biological Sciences* 264(1389):1837–1840.

Hastings, A., C. L. Hom, S. Ellner, P. Turchin, and H.C.J. Godfray. 1993. "Chaos in ecology: is mother nature a strange attractor?" *Annual Review of Ecology and Systematics* 24(1):1–33. http://arjournals.annualreviews.org/doi/abs/10.1146/annurev.es.24.110193.

Hastings, A., and T. Powell. 1991. "Chaos in a 3-species food-chain." *Ecology* 72(3):896–903.

Hastings, A., and D. Wollkind. 1982. "Age structure in predator-prey systems. 1. A general model and a specific example." *Theoretical Population Biology* 21(1):44–56.

Haydon, D. 1994. "Pivotal assumptions determining the relationship between stability and complexity: an analytical synthesis of the stability-complexity debate." *American Naturalist* 144(1):14–29.

Haydon, D. T. 2000. "Maximally stable model ecosystems can be highly connected." *Ecology* 81(9):2631–2636.

Hebert, C. E., D.V.C. Weseloh, A. Idrissi, M. T. Arts, R. O'Gorman, O. T. Gorman, B. Locke, C. P. Madenjian, and E. F. Roseman. 2008. "Restoring piscivorous fish populations in the Laurentian Great Lakes causes seabird dietary change." *Ecology* 89(4):891–897.

Holling, C. S. 1996. "Engineering versus ecological resilience. In: *Engineering with ecological constraints*, edited by P. C. Schulz, 31–43. National Academic Press, Washington, DC.

Holt, R. D. 1997. "Community modules." In: *Multi-trophic interactions in terrestrial ecosystems*, edited by A. C. Gange and V. K. Brown, 333–339. Blackwell, London.

Holt, R. D., and J. H. Lawton. 1994. "The ecological consequences of shared natural enemies," *Annual Review of Ecology and Systematics* 25:495–520.

Holt, R. D., and M. Loreau. 2001. "Biodiversity and ecosystem functioning: the role of trophic interactions and the importance of system openness." In: *The Functional consequences of biodiversity*, edited by A. P. Kinzig, S. W. Pacala, and D. Tilman, 246–262. Princeton University Press, Princeton, NJ.

Holt, R. D., and G. A. Polis. 1997. "A theoretical framework for intraguild predation." *American Naturalist* 149(4):745–764.

Holyoak, M., and S. Sachdev. 1998. "Omnivory and the stability of simple food webs," *Oecologia* 117:413–419.

Houlahan, J. E., D. J. Currie, K. Cottenie, G. S. Cumming, S.K.M. Ernest, C. S. Findlay, S. D. Fuhlendorf, U. Gaedke, P. Legendre, J. J. Magnuson, B. H. McArdle, E. H. Muldavin, D. Noble, R. Russell, R. D. Stevens, T. J. Willis, I. P. Wolwod, and S. M. Wondzell. 2007. "Compensatory dynamics are rare in natural ecological communities," *Proceedings of the National Academy of Sciences* 104(9):32–73.

Hsieh, C., C. S. Reiss, J. R. Hunter, J. R. Beddington, R. M. May, and G. Sugihara. 2006. "Fishing elevates variability in the abundance of exploited species." *Nature*, 443(7113):859–862. http://dx.doi.org/10.1038/nature05232.

Huffaker, C. B. 1958. "Experimental studies on predation: dispersion factors and predator-prey oscillations," *Hilgardia* 27:343–383.

Ives, A. R., and B. J. Cardinale. 2004. "Food-web interactions govern the resistance of communities after non-random extinctions." *Nature* 429:174–177.

Ives, A. R., and S. R. Carpenter. 2007. "Stability and diversity of ecosystems." *Science* 317(5834):58–62.

Ives, A. R., A. Einarsson, V.A.A. Jansen, and A. Gardarsson. 2008. "High-amplitude fluctuations and alternative dynamical states of midges in Lake Myvatn." *Nature*, 452(7183):84–87.

Jackson, J.B.C. 2001. "What was natural in the coastal oceans?," *Proceedings of the National Academy of Sciences of the United States of America* 98(10): 5411–5418.

Jackson, J.B.C., M. X. Kirby, W. H. Berger, K. A. Bjorndal, L. W. Botsford, B. J. Bourque, R. H. Bradbury, R. Cooke, J. Erlandson, J. A. Estes, T. P. Hughes, S. Kidwell, C. B. Lange, H. S. Lenihan, J. M. Pandol, C. H. Peterson, R. S. Steneck, M. J. Tegner, and R. R. Warner. 2001. "Historical overfishing and the recent collapse of costal ecosystems." *Science* 293:692–638.

Jefferies, R. L., A. P. Jano, and K. F. Abraham. 2006. "A biotic agent promotes large-scale catastrophic change in the coastal marshes of Hudson Bay." *Journal of Ecology* 94(1):234–242.

Jefferies, R. L., D. R. Klein, and G. R. Shaver. 1994. "Vertebrate herbivores and northern plant communities—reciprocal influences and responses." *Oikos* 71(2):193–206.

Jefferies, R. L., R. F. Rockwell, and K. F. Abraham. 2004. "Agricultural food subsidies, migratory connectivity and large-scale disturbance in arctic coastal systems: a case study." *Integrative Comparative Biology* 44:130–139.

Jennings, S., S.P.R. Greenstreet, L. Hill, G. J. Piet, J. K. Pinnegar, and K. J. Warr. 2002a. "Long-term trends in the trophic structure of the North Sea fish community:

evidence from stable-isotope analysis, size-spectra and community metrics." *Marine Biology* 141(6):1085–1097.

Jennings, S., S.P.R. Greenstreet, and J. D. Reynolds. 1999. "Structural change in an exploited fish community: a consequence of differential fishing effects on species with contrasting life histories," *Journal of Animal Ecology* 68:617–627.

Jennings, S., J. K. Pinnegar, N.V.C. Polunin, and K. J. Warr. 2002b. "Linking size-based and trophic analyses of benthic community structure." *Marine Ecology Progress Series* 226:77–85.

Jensen, A. L. 1994. "Larkin's predation model of lake trout (*Salvelinus namaycush*) extinction with harvesting and sea lamprey (*Petromyzon marinus*) predation— a qualitative analysis." *Canadian Journal of Fisheries and Aquatic Sciences* 51:942–945.

Jiang, L., H. Joshi, and S. N. Patel. 2009. "Predation alters relationships between biodiversity and temporal stability," *American Naturalist* 173:389–399.

Jonsson, T., J. E. Cohen, and S. R. Carpenter. 2005. "Food webs, body size, and species abundance in ecological community description," *Advances in Ecological Research* 36:1–84.

Jordan, C. F., J. R. Kline, and D. S. Sasscer. 1972. "Relative stability of nutrient cycles in forest ecosystems." *American Naturalist* 106:237–253.

Kaji, K., H. Takahashi, K. Hideaki, M. Kohira, and M. Yamanaka. 2009. *Sika deer: biology and management of native and introduced populations.* Springer, New York.

Kendall, B., J. Prendergast, and O. Bjornstad. 1998. "The macroecology of population dynamics: taxonomic and biogeographic patterns in population cycles." *Ecology Letters* 1:160–164.

Kolasa, J., and S.T.A. Pickett. 1989. "Ecological systems and the concept of biological organization." *Proceedings of the National Academy of Sciences of the United States of America* 86(22):8837–8841.

Kondoh, M. 2003. "Foraging adaptation and the relationship between food-web complexity and stability." *Science* 299(5611):1388–1391.

———. 2008. "Building trophic modules into a persistent food web." *Proceedings of the National Academy of Sciences of the United States of America* 105(43):16631–16635.

Krause, A. E., K. A. Frank, D. M. Mason, R. E. Ulanowicz, and W. W. Taylor. 2003. "Compartments revealed in food-web structure." *Nature* 426(6964):282–285.

Krivan, V., and O. J. Schmitz. 2004. "Trait and density mediated indirect interactions in simple food webs." *Oikos* 107(2):239–250.

Kuznetsov, Y. A., and S. Rinaldi. 1996. "Remarks on food chain dynamics." *Mathematical Biosciences* 134(1):1–33. http://www.sciencedirect.com/science/article/B6VHX-3WJESVP-1/2/9f754f68a62d71C8d8907d14aa26f7a0.

Laska, M. S., and J. T. Wootton. 1998. "Theoretical concepts and empirical approaches to measuring interaction strength." *Ecology* 79(2):461–476.

Leeuwen, A. V., A. De Roos, and L. Persson. 2008. "How cod shapes its world." *Journal of Sea Research* 60:89–104.

Leroux, S. J., and M. Loreau. 2008. "Subsidy hypothesis and strength of trophic cascades across ecosystems." *Ecology Letters* 11(11):1147–1156.

Levin, S. 1999. *Fragile dominion: complexity and the commons.* Perseus, Santa Fe, NM.

Levin, S. A. 1992. "The problem of Pattern and Scale in Ecology." *Ecology* 73(6): 1943–1967.

———. 2003. "Complex adaptive systems: exploring the known, the unknown and the unknowable." *Bulletin of the American Mathematical Society* 40(1):319.

———. 2005. "Self-organization and the emergence of complexity in ecological systems," *Bioscience* 55(12):1075–1079.

Levins, R. 1968. *Evolution in changing environments*. Princeton University Press, Princeton, NJ.

———. 1975. "Evolution in communities near equilibrium." In: *Ecology and evolution of communities*, edited by M. L. Cady and J. M. Diamond, 16–50. Belknap Press, Boston.

Loeuille, N., and M. Loreau. 2005. "Evolutionary emergence of size-structured food webs." *Proceedings of the National Academy of Sciences of the United States of America* 102(16):5761–5766. http://www.pnas.org/content/102/16/5761.abstract.

Logan, J. A., J. Regniere, and J. A. Powell. 2003. "Assessing the impacts of global warming on forest pest dynamics." *Frontiers in ecology and the environment* 1(3):130–137.

Loreau, M. 1994. "Material cycling and the stability of ecosystems." *American Naturalist* 143(3):508–513.

Luckinbill, L. S. 1973. "Coexistence in laboratory populations of *Paramecium aurelia* and its predator *Didinium nasutum*." *Ecology* 54:1320–1327.

MacArthur, R. H., and E. R. Pianka. 1966. "On the optimal use of a patchy environment." *American Naturalist* 100:603–609.

MacArthur, R. H., and E. O. Wilson. 1967. *The theory of island biogeography*, Princeton University Press, Princeton, NJ.

Martin, N. V. 1966. The significance of food habits in the biology, exploitation, and management of Algonquin Park, Ontario, lake trout." *Transactions of the American Fisheries Society* 95(4):415–422.

Martinez, N. D. 1991a. "Artifacts or attributes—effects of resolution on the Little Rock Lake food web." *Ecological Monographs* 61(4):367–392.

———. 1993. "Effect of Scale on Food Web structure." *Science* 260(5105):242–243.

Martinez, N. D. 1991. "Artifacts or attributes? Effects of resolution on the Little Rock Lake food web." *Ecological Monographs* 61:367–392.

Martinez, N. D., R. J. Williams, and J. A. Dunne. 2006. "Diversity, complexity and persistence in large model ecosystems." In: *Ecological networks: linking structure to dynamics in food webs* edited by M. Pascual and J. A. Dunne, 46–57. Oxford University Press, Oxford.

Maser, G. L., F. Guichard, and K. S. McCann. 2007. "Weak trophic interactions and the balance of enriched metacommunities." *Journal of Theoretical Biology* 247(2):337–345.

May, R. M. 1973. "Relationships among various types of population models." *American Naturalist* 107(953): 46–57.

———. 1974a. "Biological populations with nonoverlapping generations—stable points, stable cycles, and chaos." *Science* 186(4164):645–647.

———. 1974b. "On the theory of Niche Overlap." *Theoretical Population Biology* 5(3):297–332.

———. 1976. "Simple mathematical models with very complicated dynamics." *Nature* 261:459–467.

McCann, K. 1998. "Density-dependent coexistence in fish communities." *Ecology* 79(8):2957–2967. http://www.jstor.org/stable/176529.

McCann, K., and A. Hastings, 1997. "Re-evaluating the omnivory-stability relationship in food webs." *Proceedings of the Royal Society of London Series B: Biological Sciences.* 264(1385):1249–1254.

McCann, K., A. Hastings, S. Harrison, and W. Wilson. 2000. "Population outbreaks in a discrete world." *Theoretical Population Biology* 57(2): 97–108.

McCann, K., A. Hastings, and G. R. Huxel. 1998. "Weak trophic interactions and the balance of nature." *Nature* 395(6704):794–798.

McCann, K. S., J. B. Rasmussen, and J. Umbanhowar. 2005. "The dynamics of spatially coupled food webs." *Ecology Letters* 8(5):513–523.

McCann, K. S., and N. Rooney. 2009. "The more food webs change, the more they stay the same." *Proceedings of the Royal Society of London Series B: Biological Sciences* 364:1789–1801.

McCann, K., and P. Yodzis. 1994. "Biological conditions for chaos in a three-species food chain." *Ecology* 75(2):561–564. http://www.jstor.org/stable/1939558.

———. 1995. "Bifurcation structure of a 3-species food-chain model." *Theoretical Population Biology* 48(2):93–125.

———. 1998. "On the stabilizing role of stage structure in piscene consumer-resource interactions." *Theoretical Population Biology* 54(3):227–242.

McCauley, E., W. A. Nelson, and R. M. Nisbet. 2008. "Small-amplitude cycles emerge from stage- structured interactions in daphnia-algal systems." *Nature* 455:1240–1243.

McCauley, E., R. M. Nisbet, W. W. Murdoch, A. M. De Roos, and W.S.C. Gurney. 1999. "Large-amplitude cycles of daphnia and its algal prey in enriched environments." *Nature* 402(6762):653–656.

McCauley, E., W. Wilson, and A. M. De Roos. 1996. "Dynamics of age-structured predator-prey populations in space: asymmetrical effects of mobility in juvenile and adult predators." *Oikos* 76:485–497.

McCauley, E., W. G. Wilson, and A. M. De Roos. 1993. "Dynamics of age-structured and spatially structured predator-prey interactions—individual-based models and population-level formulations." *American Naturalist* 142(3):412–442.

Menge, B. A. 1995. "Indirect effects in marine rocky intertidal interaction webs—patterns and importance." *Ecological Monographs* 65(1):21–74.

Milo, R., S. Shen-Orr, S. Itzkovitz, N. Kashtan, D. Chklovskii, and U. Alon. 2002. "Network motifs: simple building blocks of complex networks." *Science* 298(5594):824–827.

Montoya, J. M., S. L. Pimm, and R. V. Sole. 2006. "Ecological networks and their fragility." *Nature* 442(7100):259–264.

Moore, J. C., E. L. Berlow, D. C. Coleman, P. C. de Ruiter, Q. Dong, A. Hastings, N. C. Johnson, K. S. McCann, K. Melville, P. J. Morin, K. Nadelhoffer, A. D. Rosemond, D. M. Post, J. L. Sabo, K. M. Scow, M. J. Vanni, and D. H. Wall. 2004. "Detritus, trophic dynamics and biodiversity." *Ecology Letters* 7(7): 584–600.

Moore, J. C., and H. W. Hunt. 1988. "Resource compartmentation and the stability of real ecosystems." *Nature* 333(6170):261–263.

Morbey, Y., P. Addison, B. J. Shuter, and K. Vascotto. 2006. "Within population heterogeneity of habitat use by lake trout *Salvelinus namaycush.*" *Journal of Fish Biology* 69:1675–1679.

Muratori, S., and S. Rinaldi. 1992. "Low- and high-frequency oscillations in three-dimensional food chian systems." *SIAM Journal of Applied Mathematics* 52:1688–1706.

Murdoch, W. M., C. J. Briggs, and Nisbet, R. M. 2003. *Consumer-resource dynamics*, Princeton University Press, Princeton, NJ.

Murdoch, W. W., B. E. Kendall, R. M. Nisbet, C. J. Briggs, E. McCauley, and R. Bolser, 2002. "Single-species models for many-species food webs." *Nature* 417(6888):541–543.

Myers, R. A. 2001. "Stock and recruitment: generalizations about maximum reproductive rate, density dependence, and variability using meta-analytic approaches." *ICES Journal of Marine Science* 58(5):937.

Myers, R. A., and B. Worm. 2003. "Rapid worldwide depletion of predatory fish communities." *Nature* 423(6937):280–283.

Nakajima, H., and D. L. DeAngelis. 1989. "Resilience and local stability in a nutrient-limited resource-consumer system." *Bulletin of Mathematical Biology* 51:501–510.

Neubert, M., and M. Kot. 1992. "The subcritical collapse of predator populations in discrete-time predator-prey models." *Mathematical Biosciences* 110:45–66.

Neutel, A. M., J.A.P. Heesterbeek, and P. C. de Ruiter. 2002. "Stability in real food webs: weak links in long loops." *Science* 296(5570):1120–1123.

Neutel, A. M., J.A.P. Heesterbeek, J. van de Koppel, G. Hoenderboom, A. Vos, C. Kaldeway, F. Berendse, and P. C. de Ruiter. 2007. "Reconciling complexity with stability in naturally assembling food webs." *Nature* 499:599–602.

Nicholson, A. J., and V. A. Bailey. 1935. "The balance of animal populations." *Proceedings of the Zoological Society of London* 3:551–598.

O'Gorman, E., R. Enright, and M. Emmerson. 2008. "Predator diversity enhances secondary production and decreases the likelihood of trophic cascades." *Oecologia* 158(3):557–567.

Oksanen, L., S. D. Fretwell, J. Arruda, and P. Niemela. 1981. "Exploitation ecosystems in gradients of primary productivity." *American Naturalist* 118(2):240. http://www.journals.uchicago.edu/doi/abs/10.1086/283817.

O'Neill, R. V. 1976. "Ecosystem persistence and heterotrophic regulation." *Ecology* 57(6):1244–1253.

Otto, S. B., B. C. Rall, and U. Brose. 2007. "Allometric degree distributions facilitate food-web stability." *Nature* 450:1226–1230.

Owen-Smith, N., and M.G.L. Mills. 2008. "Shifting prey selection generates contrasting herbivore dynamics within a large-mammal predator-prey web." *Ecology* 89(4):1120–1133.

Paine, R. T. 1980. "Food webs: linkage, interaction strength and community infrastructure." *Journal of Animal Ecology* 49(3):667–685.

Paine, R. T. 1992. "Food-web analysis through field measurement of per-capita interaction strength." *Nature* 355(6355):73–75.

Pauly, D., V. Christensen, J. Dalsgaard, R. Froese, and F. Torres. 1998. "Fishing down marine food webs." *Science* 279(5352):860–863.

Persson, L., P. A. Amundsen, A. M. De Roos, A. Klemetsen, R. Knudsen, and R. Primicerio. 2007. "Culling prey promotes predator recovery—alternative states in a whole-lake experiment." *Science* 316(5832):1743–1746.

Persson, L., and A. M. de Roos. 2003. "Adaptive habitat use in size-structured populations: linking individual behavior to population processes." *Ecology* 84(5):1129–1139.

Petchey, O. L., A. P. Beckerman, J. O. Riede, and P. H. Warren. 2008. "Size, foraging, and food web structure." *Proceedings of the National Academy of Sciences of the United States of America* 105(11):4191–4196.

Peters, R. H. 1983. *The ecological implications of body size.* Cambridge University Press, New York.

Pimm, S. L. 1979. "The structure of food webs." *Theoretical Population Biology* 16(2):144–158.

———. 1982. *Food webs.* University of Chicago Press.

———. 1984. "The Complexity and stability of Ecosystems." *Nature,* 307(5949): 321–326.

———. 1987. "Determining the effects of introduced species." *Trends in Ecology & Evolution* 2(4):106–108.

Pimm, S. L., and R. L. Kitching. 1987. "The determinants of food-chain lengths." *Oikos* 50(3):302–307.

Pimm, S. L., and J. Lawton. 1978. "On feeding on more than one trophic level." *Nature* 275:542–544.

———. 1980. "Are food webs divided into compartments?" *Journal of Animal Ecology* 49(3):879–898.

Pimm, S. L., J. H. Lawton, and J. E. Cohen, "Food web patterns and their consequences." *Nature* 350(6320):669–674.

Polis, G. A. 1991. "Complex trophic interactions in deserts—an empirical critique of food-web theory." *American Naturalist* 138(1):123–155.

Polis, G. A., W. B. Anderson, and R. D. Holt. 1997. "Toward an integration of landscape and food web ecology: the dynamics of spatially subsidized food webs." *Annual Review of Ecology and Systematics* 28:289–316.

Polis, G. A., and R. D. Holt. 1992. "Intraguild predation—the dynamics of Complex Trophic Interactions." *Trends in Ecology & Evolution* 7(5):151–154.

Polis, G. A., R. D. Holt, B. A. Menge, and K. O. Winemiller. 1996. "Time, space, and life history: influences on food webs." In: *Food webs: integration of patterns and dynamics,* edited by G. A. Polis and K. Winemiller, 435–460. Chapman & Hall, London.

Polis, G. A., and S. D. Hurd. 1996. "Linking marine and terrestrial food webs: allochthonous input from the ocean supports high secondary productivity on small islands and coastal land communities." *American Naturalist* 147(3): 396–423.

Polis, G. A., and D. R. Strong. 1996. "Food web complexity and community dynamics." *American Naturalist* 147(5):813–846.

Post, D. M. 2002. "The long and short of food-chain length." *Trends in Ecology & Evolution* 17(6):269–277.

Post, D. M., M. L. Pace, and N. G. Hairston. 2000. "Ecosystem size determines food-chain length in lakes." *Nature* 405(6790):1047–1049.

Post, D. M., and G. Takimoto. 2007. "Proximate structural mechanisms for variation in food-chain length." *Oikos* 116(5):775–782.

Raffaelli, D. 2002. "Ecology—from Elton to mathematics and back again." *Science* 296(5570):1035–1037.

Raffaelli, D., and S. J. Hall. 1992. "Compartments and predation in an estuarine food web." *Journal of Animal Ecology* 61(3):551–560.

Rall, B. C., C. Guill, and U. Brose. 2008. "Food-web connectance and predator interference dampen the paradox of enrichment." *Oikos* 117(2):202–213.

Rasmussen, J. B., D. J. Rowan, D.R.S. Lean, and J. H. Carey. 1990. "Food chain structure in Ontario lakes determines PCB levels in lake trout (*Salvelinus namaycush*) and other pelagic fish." *Canadian Journal of Fisheries and Aquatic Sciences* 47(10):2030–2038.

Reuman, D. C., and J. E. Cohen. 2004. "Trophic links' length and slope in the Tuesday Lake food web with species' body mass and numerical abundance." *Journal of Animal Ecology* 73(5):852–866.

Ricker, W. E. 1954. "Stock and recruitment." *Journal of the Fisheries Research Board* 11:559–623.

Rip, J., K. S. McCann, D. H. Lynn, and S. Fawcett. 2010. "An experimental test of a fundamental food web motif." *Proceedings of the Royal Society of London Series B: Biological Sciences* 277:1743–1749.

Ripple, W. J., E. J. Larsen, R. A. Renkin, and D. W. Smith. 2001. "Trophic cascades among wolves, elk and aspen on Yellowstone National Park's northern range." *Biological Conservation* 102(3):227–234.

Romanuk, T. N., B. E. Beisner, N. D. Martinez, and J. Kolasa. 2006. "Non-omnivorous generality promotes population stability." *Biology Letters* 2(3):374–377.

Rooney, N., K. McCann, G. Gellner, and J. C. Moore. 2006. "Structural asymmetry and the stability of diverse food webs." *Nature* 442(7100):265–269.

Rooney, N., K. S. McCann, and J. C. Moore. 2008. "A landscape theory for food web architecture." *Ecology Letters* 11(8):867–881.

Rosenzweig, M. L. 1971. "Paradox of enrichment: destabilization of exploitation ecosystems in ecological time." *Science* 171(3969): 385–387.

Rosenzweig, M. L. 1995. *Species diversity in space and time*. Cambridge University Press. Cambridge.

Rosenzweig, M. L., and R. H. MacArthur. 1963. "Graphical representation and stability conditions of predator-prey interactions." *American Naturalist* 97:895.

Rudolf, V.H.W. 2008. "Consequences of size structure in the prey for predator:prey dynamics: the composite functional response." *Journal of Animal Ecology* 77(3):520–528. http://dx.doi.org/10.1111/j.1365-2656.2008.01368.x.

Scheffer, M., S. Carpenter, J. A. Foley, C. Folke, and B. Walker. 2001. "Catastrophic shifts in ecosystems." *Nature* 413(6856):591–596.

Schmitz, O. J., P. A. Hamback, and A. P. Beckerman. 2000. "Trophic cascades in terrestrial systems: a review of the effects of carnivore removals on plants." *American Naturalist* 155(2):141–153.

Schoener, T. W. 1989. "Food webs from the small to the large." *Ecology* 70:1559–1589.

Schreiber, S. J. 2001. "Chaos and population disappearances in simple ecological models." *Journal of Mathematical Biology* 42(3): 239–260.

Schreiber, S., and V. W. Rudolf. 2008. "Crossing habitat boundaries: coupling dynamics of ecosystems through complex life cycles." *Ecology Letters* 11:576–587.

Schwartz, M. K., L. S. Mills, K. S. McKelvey, L. F. Ruggiero, and F. W. Allendorf. 2002. "DNA reveals high dispersal synchronizing the population dynamics of Canada lynx." *Nature* 415(6871):520–522.

Sherr, E. B., and B. F. Sherr. 1991. "Planktonic microbes—tiny cells at the base of the oceans' food webs." *Trends in Ecology & Evolution* 6(2): 50–54.

Sherwood, G. D., J. Kovecses, A. Hontela, and J. B. Rasmussen. 2002a. "Simplified food webs lead to energetic bottlenecks in polluted lakes." *Canadian Journal of Fisheries and Aquatic Sciences* 59(1):1–5.

Sherwood, G. D., I. Pazzia, A. Moeser, A. Hontela, and J. B. Rasmussen. 2002b. "Shifting gears: enzymatic evidence for the energetic advantage of switching diet in wild-living fish." *Canadian Journal of Fisheries and Aquatic Sciences* 59(2):229–241.

Shurin, J. B., E. T. Borer, E. W. Seabloom, K. Anderson, C. A. Blanchette, B. Broitman, S. Cooper, and B. S. Halpern. 2002. "A cross-ecosystem comparison of the strength of trophic cascades." *Ecology Letters* 5:785–791.

Shuter, B. J., M. L. Jones, and R. M. Korver. 1998. "A general, life history based model for regional management of fish stocks: the inland lake trout (*Salvelinus namaycush*) fisheries of Ontario." *Canadian Journal of Fisheries and Aquatic Sciences* 55:2161–2177.

Smith, D. R. "Change and variability in Climate and Ecosystem Decline in Aral Sea Basin Deltas." *Post-Soviet Geography* 35(3):142–165.

Solow, A. R., and A. Beet. 1998. "On lumping species in food webs." *Ecology* 79: 2013–2018.

Solow, A. R., C. Costello, and A. Beet. 1999. "On an early result on stability and complexity." *American Naturalist* 154(5):587–588.

Sommer, U., Z. M. Gliwicz, W. Lampert, and A. Duncan. 1986. "The peg-model of seasonal succession of planktonic events in fresh waters." *Archiv für Hydrobiologie*. 106(4):433–471.

Sprules, W. G., and J. E. Bowerman. 1988. "Omnivory and food chain length in zooplankton food webs." *Ecology* 69(2):418–426. http://www.esajournals. org/doi/abs/10.2307/1940440.

Srivastava, D. S., and R. L. Jefferies. 1996. "A positive feedback: herbivory, plant growth, salinity, and the desertification of an Arctic salt-marsh." *Journal of Ecology* 84(1):31–42.

Stapp, P., and G. A. Polis. 2003. "Influence of pulsed resources and marine subsidies on insular rodent populations." *Oikos* 102(1):111–123.

Stouffer, D. B. 2005. "Quantitative patterns in the structure of model and empirical food webs." *Ecology* 86:1301–1311.

Stouffer, D. B., and L.A.N. Amaral. 2007. "Evidence for the existence of a robust pattern of prey selection in food webs." *Proceedings of the Royal Society of London Series B: Biological Sciences* 274:1931–1940.

Strickler, K. 1979. "Specialization and foraging efficiency of solitary bees." *Ecology* 60:998–1009.

Strong, D. R. 1992. "Are trophic cascades all wet—differentiation and donor control in speciose ecosystems." *Ecology* 73(3):747–754.

Sugihara, G., K. Schoenly, and A. Trombla. 1989. "Scale invariance in food web properties." *Science* 245(4913):48–52. http://www.sciencemag.org/cgi/content/abstract/245/4913/48.

Tanabe, K., and N. Toshiyuki. 2005. "Omnivory creates chaos in simple food web models." *Ecology* 86:3411–3414.

Tanner, J. T. 1975. "The stability and the intrinsic growth rates of prey and predator populations." *Ecology* 56:855–867.

Tegner, M. J., and P. K. Dayton. 2000. "Ecosystem effects of fishing in kelp forest communities." *ICES Journal of Marine Science* 57(3):579–589.

Teng, J., and K. S. McCann. 2004. "Dynamics of compartmented and reticulate food webs in relation to energetic flows." *American Naturalist* 164(1):85–100.

Terborgh, J., L. Lopez, P. Nunez, M. Rao, G. Shahabuddin, G. Orihuela, M. Riveros, R. Ascanio, G. H. Adler, T. D. Lambert, and L. Balbas. 2001. "Ecological meltdown in predator-free forest fragments." *Science* 294(5548):1923–1926.

Tewfik, A., F. Guichard, and K. S. McCann. 2007a. "Influence of acute and chronic disturbance on macrophyte landscape zonation." *Marine Ecology Progress Series* 335:111–121.

Tewfik, A., J. B. Rasmussen, and K. S. McCann. 2007b. "Simplification of seagrass food webs across a gradient of nutrient enrichment." *Canadian Journal of Fisheries and Aquatic Sciences* 64:956–967.

Thompson, R. M., M. Hemberg, B. M. Starzomski, and J. B. Shurin. 2007. "Trophic levels and trophic tangles: the prevalence of omnivory in real food webs." *Ecology* 88(3):612–617.

Tilman, D. 1982. *Resource competition and community structure*, Princeton University Press, Princeton, NJ.

———. 1996. "Biodiversity: population versus ecosystem stability." *Ecology* 77(2):350–363.

Tilman, D., C. L. Lehman, and C. E. Bristow. 1998. "Diversity-stability relationships: statistical inevitability or ecological consequence?" *American Naturalist* 151(3):277–282.

Tilman, D., D. Wedin, and J. Knops. 1996. "Productivity and sustainability influenced by biodiversity in grassland ecosystems." *Nature* 379(6567):718–720.

Tunney, T., K. S. McCann, and B. J. Shuter. Forthcoming. "Food webs as complex adaptive systems: an empirical example."

Turchin, P., and I. Hanski. 2001. "Contrasting alternative hypotheses about rodent cycles by translating them into parameterized models." *Ecology Letters* 4:267–276.

Turchin, P., S. N. Wood, S. P. Ellner, B. E. Kendall, W. W. Murdoch, A. Fischlin, J. Casas, E. McCauley, and C. J. Briggs. 2003. "Dynamical effects of plant quality and parasitism on population cycles of larch budmoth." *Ecology* 84(5):1207–1214.

Tylianakis, J. M., T. Tscharntke, and O. T. Lewis. 2007. "Habitat modification alters the structure of tropical host-parasitoid food webs." *Nature* 445(7124):202–205.

Vadeboncoeur, Y., M. J. Vander Zanden, and D. M. Lodge. 2002. "Putting the lake back together: reintegrating benthic pathways into lake food web models." *Bioscience* 52(1):44–54.

Vandermeer, J. 2006. "Omnivory and the stability of food webs." *Journal of Theoretical Biology* 238(3):497–504.

Vandermeer, J., and P. Yodzis. 1999. "Basin boundary collision as a model of discontinuous change in ecosystems." *Ecology* 80(6):1817–1827.

Vander Zanden, M. J., and J. B. Rasmussen. 1999. "Primary consumer delta^{13}C and delta^{15}N and the trophic position of aquatic consumers." *Ecology (Washington DC)* 80(4):1395–1404.

Vander Zanden, M. J., and J. B. Rasmussen. 2001. "Variation in delta N-15 and delta C-13 trophic fractionation: implications for aquatic food web studies." *Limnology and Oceanography* 46(8):2061–2066.

Vasseur, D. A., U. Gaedke, and K. S. McCann. 2005. "A seasonal alternation of coherent and compensatory dynamics occurs in phytoplankton." *Oikos* 110(3):507–514.

Walters, C. J., and J. F. Kitchell. 2001. "Cultivation/depensation effects on juvenile survival and recruitment: implications for the theory of fishing." *Canadian Journal of Fisheries and Aquatic Sciences* 58:39–50.

Werner, E. E., and B. R. Anholt. 1993. "Ecological consequences of the trade-off between growth and mortality rates mediated by foraging activity." *American Naturalist* 142(2):242. http://www.journals.uchicago.edu/doi/abs/10.1086/285537. PMID: 19425978.

Wiggins, S. 1990. *Introduction to applied nonlinear dynamical systems and chaos*, Springer, Berlin.

Wilhelm, S. W., G. S. Bullerjahn, M. L. Eldridge, J. M. Rinta-Kanto, L. Poorvin, and R. A. Bourbonniere. 2006. "Seasonal hypoxia and the genetic diversity of prokaryote populations in the central basin hypolimnion of Lake Erie: evidence for abundant cyanobacteria and photosynthesis." *Journal of Great Lakes Research* 32(4):657–671.

Williams, R. J. 2000. "Simple rules yield complex food webs." *Nature* 404:180–183.

Williams, R. J. 2008. "Effects of network and dynamical model structure on species persistence in large model food webs." *Theoretical Ecology* 1(3):141–151.

Wilson, E. O. 1992. *The diversity of life*. Harvard University Press, Cambridge, MA.

Winemiller, K. O. 1990. "Spatial and temporal variation in tropical fish trophic networks." *Ecological Monographs* 60(3):331–367.

Woodward, G., and A. G. Hildrew. 2002. "Food web structure in riverine landscapes." *Freshwater Biology* 47(4):777–798.

Woodward, G., D. C. Speirs, and A. G. Hildrew. 2005. "Quantification and resolution of a complex, size-structured food web." In: *Food webs: from connectivity to energetics*, edited by L. Yigi, 36:85–131. Elsevier, New York.

Wootton, J. T. 2005. "Field parameterization and experimental test of the neutral theory of biodiversity." *Nature* 433(7023):309–312.

Yodzis, P. 1981. "The stability of real ecosystems." *Nature* 289(5799):674–676.

———. 1988. "The indeterminacy of ecological interactions as perceived through perturbation experiments." *Ecology* 69(2):508–515.

———. 1989. *Introduction to theoretical ecology*. Harper-Collins, New York.

———. 1994. "Predator-prey theory and management of multispecies fisheries." *Ecological Applications* 4(1):51–58.

Yodzis, P., and S. Innes. 1992. "Body size and consumer-resource dynamics." *American Naturalist* 139(6):1151–1175.

Index

adaptive behavior, 127–132, 151, 164, 201–217

age structure. *See* stage structure

agriculture, 10, 17, 214

algae/algal, 13, 42–44, 84, 100–101, 137, 151, 166–167

Algonquin Park, 208

allochthonous inputs, subsidies, 11–13, 17, 213–215

allometry, 85, 121, 161–162, 187

alternative states: biological example of, 16; definition of, 8, 9; mathematical discussion of, 26; with stage structure, 97–102

antipredator, antiherbivore defenses, 13–14, 85, 133, 165

apparent competition, 48, 95, 103, 123–124, 134–140

aquatic ecosystems: contrasted with terrestrial ecosystems, 85–87, 151, 162; food web structure of, 147–151, 160–162; microcosms, 42, 84, 100, 137; omnivory in, 119; shape, 204, 207–211; size, 158–161, 207–212

Aral Sea, 16

asymmetric interactions, 128–130, 137, 141, 160, 177

asynchrony: generation of, 167–169; and stability, 157, 163–169

bacteria, 6, 13, 149, 180

bacteriovores, 119

behavior. *See* adaptive behavior

bifurcation, 20–22, definition of, 21; as experiments, 35–39, 45; biological example of, 42–45; Hopf, 73–74, 190, 196–197; transcritical, 107, of food chain, 113–117, of modules, 137, 140, 156; approach and community matrices, 171; in ecosystem models, 190, 196–197, 198

biomass pyramid, 76–77, 86, 157–160, 178, 204–215; Eltonian definition of, 77; inverted or wasp-wasted definition of, 77

body-size, 85, 118, 121, 146–151, 161–162, 169, 185–188

boreal forests, 164

boreallakes, 213

chaos, 24, 27, 91–93, 109, 113–114, 133–136

characteristic polynomial or equation, 32, 62, 175, 178

cisco, 97, 206–213

coastal ecosystems, 13, 15, 187, 214

coefficient of variation, 4, 8, 44–45, 86, 107, 119–120, 186, 217

community matrix, 32, 68, 170–175, 178–180, 184–185, 188

compartments: detrital, 5, 191, patterns of body-size in, 161; spatial or habitat, 147, 151, 181; in whole communities, 171, 176–183, 188

competition, 47–48, 99, 124–126, 131, 134–140

coexistence, 121, 125, 127, 136

cohort cycles. *See* generation cycles

collapse, 6, 10–13, 15–17, 38, 105, 159, 217

connectance, 125, 182

continuous equations or continuous time, 27, 30, 45, 53–62, 67, 79, 89–91, 166

coupling terms or coupling strength, 70, 73, 77–88, 94, 101–102, 105–112, 117, 121, 124, 132–135, 172, 177; due to habitat coupling or mobile food web consumers, 145–169; due to ontogenetic habitat, 89; of whole communities, 182–188

culling. *See* harvesting

cycles. *See* oscillation

dead zone, 13–15

DeAngelis, D.: on diversity-stability, 185; on ecosystem resilience, 190, 194–195

deer, 64, 216–217

desertification, 16

desert ecosystems, 213–214

detritus, 5, 13–14, 149, 189–200, 214

MONOGRAPHS IN POPULATION BIOLOGY
EDITED BY SIMON A. LEVIN AND HENRY S. HORN